ГЕОЛОГИЧЕСКАЯ ДЕЯТЕЛЬНОСТЬ МИКРООРГАНИЗМОВ

GEOLOGICHESKAYA DEYATEL'NOST' MIKROORGANIZMOV

GEOLOGIC ACTIVITY OF MICROORGANISMS

Transactions (Trudy) of the Institute of Microbiology No. IX

GEOLOGIC ACTIVITY OF MICROORGANISMS

S. I. Kuznetsov
Editor

Authorized translation from the Russian

Springer Science+Business Media, LLC 1962

ISBN 978-1-4899-4598-3   ISBN 978-1-4899-4596-9 (eBook)
DOI 10.1007/978-1-4899-4596-9

The Russian text was published by the USSR Academy of Science Press in Moscow in 1961

Труды Института микробиологии, вып. IX

Геологическая деятельность микроорганизмов

Library of Congress Catalog Card Number 62-12850

© 1962 Springer Science+Business Media New York
Originally published by Consultants Bureau Enterprises, Inc in 1962.
Softcover reprint of the hardcover 1st edition 1962

CONTENTS

CONTENTS

FOREWORD

The papers published in this ninth number of the Transactions of the Institute of Microbiology are devoted to the geological activity of microorganisms, to an explanation of the role of microorganisms in the formation of ore deposits, and to the use of microorganisms in prospecting for gas and oil; they represent material presented at the coordination conference that met in Moscow in December of 1959.

The calling of such a meeting attests to the fact that the time has arrived to gather together the results of investigations in this field, not only in Moscow, but in other cities as well, and to plan the course of development of such investigations.

There is no doubt that geological microbiology has had a stormy history. A. N. Nesmeyanov has noted that this is one of the aspects of growth in science which afterwards sends forth new shoots. The processes activated by microorganisms as geological agents have long been contemplated, discussed, and described, but in no way did we interfere with them, did we control the life activity of these microorganisms, did we strive to regulate these processes, to retard undesirable and accelerate useful processes.

It is now possible to see that a fusion of microbiological and geological investigations is occurring. The question of gaining molecular sulfur from formational water is raised; the beneficiation of lead-zinc ores is studied, and considerable work has already been accomplished in this direction in the U. S. A. and in the German Democratic Republic (East Germany). In Czechoslovakia, where there are no deposits of sulfur, molecular sulfur has been obtained from hydrogen-sulfide in sewage.

It should be noted that many stages in the development of geomicrobiology have already been passed, such as that concerning the sterility of the interior of the earth. Scarcely any geologist will now maintain that there are no microorganisms at great depths or will categorically deny the role of microorganisms in the formation of various fossil fuels and some ore deposits. B. V. Perfil'ev believes that the controversial discussions between geologists and microbiologists are completely at an end.

The ideas of Vernadskii, who maintained that microorganisms are powerful agents of disaggregation and concentration, of dissemination and accumulation of individual elements, have obtained full acceptance.

Many interesting discoveries in the field of geological microbiology lie ahead of us.

A. A. Imshenetskii

PRINCIPAL TRENDS IN THE INVESTIGATION
OF GEOLOGICAL ACTIVITY OF MICROORGANISMS

S. I. Kuznetsov

(Institute of Microbiology, Academy of Sciences, USSR)

Introduction

Among the geologic processes at work in the earth's crust, there are those that lead both to the formation as well as to the destruction of mineral deposits. Some of these deposits, such as coal, oil, some sulfur, iron ore, and others, were formed in the geologic past; others are forming at the present time.

These latter include accumulations of plant remains to form peat, the development and concentration of hydrogen sulfide in formational waters, and the formation of iron ores in oligotrophic lakes and the sea.

The deposits that have already formed do not remain stable, but have a life of their own, and, depending on changes in the surrounding environment, undergo diagenesis; during exploitation they may disintegrate quickly. Microorganisms may play an important role in the processes of disintegration. They can be easily used for producing oxidation both in organic and in some mineral substances. Sometimes the metabolism products of microorganisms create favorable conditions for chemical oxidizing processes, and deposits are destroyed more energetically.

In studying the geological activity of microorganisms we should distinguish present-day processes from those that terminated in the geologic past.

In both cases it is necessary to know, firstly, the physiology of the investigated group of microorganisms and, secondly, the environmental conditions in which the indicated group of organisms occurred or occurs.

In studying present-day processes of transforming substances in nature, the ecological conditions may be determined directly. To find a solution to the question of how some rocks were formed in the distant past and of what role microorganisms played in this process, we are reduced to the search for present-day analogies of such processes and to interpolate these back to that distant past (Isachenko, 1958).

A Formulation of the Questions Underlying this Study

By geological activity we mean that work of microorganisms that leads to alternation and redistribution of substances and elements in the atmosphere, hydrosphere, and lithosphere of the earth. In this context the questions related to the study of geological activity of microorganisms may be divided into two groups. On the one hand, there are questions concerning the formation, transformation, and destruction of organic substances and compounds of sulfur, phosphorus, and iron directly in an aqueous basin, and, on the other, there are the problems of diagenesis of previously formed deposits during their burial and of weathering of ore deposits, especially of sulfur, sulfide ores, and others.

Thus, this study is based on a determination of the role of microorganisms in the origin of mineral deposits, in the present-day processes of rock diagenesis (such as the oxidation of hydrothermal sulfide ores), in alterations of the saline content of deep waters, and in the change of these waters from sulfate to chloride or bicarbonate types.

The Conditions Necessary for the Geological Activity of Microorganisms in Nature

Before embarking on the study of any biological processes, we should determine the conditions under which any given biological process is effective, i.e., ascertain the presence of the given group of organisms, the presence of an altered substratum and of sources of organic material or other types of energy-bearing substance by which the given group of organisms may develop, and the proper ecological conditions of the group of organisms.

For example, if we study the process of denitrification, there should be denitrifying bacteria, assimilable organic material, anaerobic conditions and a neutral chemical environment where nitrates are present or accessible.

The presence of some denitrifying bacteria is not sufficient basis for concluding that denitrification is actually occurring.

The formation of hydrogen sulfide by reduction of sulfates may occur only when all the following are present: desulfurizing bacteria, sulfates, organic material or molecular hydrogen, an rH_2 below 10, and pH of about 7.

If a single one of these conditions is not met, one cannot conclude positively that the given process actually is in effect.

The Basic Plan for Studying the Geological Activity of Microorganisms

Having placed before ourselves the task of studying some process in which, in our view, the activity of a definite group of microorganisms may be important, we should, on the basis of our knowledge of the literature, formulate a working hypothesis and lay out a plan of investigation.

It is necessary to have all information on the geologic structure of the given deposit and on the hydrogeology and chemical composition of the water and the dissolved gases in order to gain a complete picture of the ecological environment in which the microorganisms act.

Further, we should make qualitative and quantitative computations of those groups of microorganisms that may play a fundamental role in the given process. A single occurrence of a given group of organisms indicates merely a possible role in the geologic process.

A more definite conclusion may be reached, after the organisms that are present have been compared with the ecological conditions and with analyses of materials in the substratum, which is basic for the given process.

Thus, the presence of putrefying bacteria in sediments of the Black Sea, which contain practically no albuminous substances, does not give grounds for concluding, as Kriss has done, that the hydrogen sulfide in the deep waters of the Black Sea formed by putrefaction of albumens.

In their studies of the formation of hydrogen sulfide in sediments of the Black Sea, Kriss and Rukina (1949) compared the intensities of hydrogen-sulfide formation on Baars medium and on a medium for putrefying bacteria to which 26 g/liter of albumen was added in order that the initial quantity of sulfatic sulfur in Baars medium and in the albumen medium for the putrefying bacteria would be the same. All the experiments on these media showed a greater production of hydrogen sulfide on the medium with albumen than on Baars medium. From these experiments the authors concluded that the hydrogen sulfide in the Black Sea forms by putrefying processes.

In this circumstance Bruevich (1953) is completely correct in his assertion that it is not at all a matter of what bacteria are present in the ground in a latent state, but rather of what function under the actual prevailing conditions the bacteria can perform.

Thus, it is not enough to find microorganisms that are able to decompose albumen and give off hydrogen sulfide, it is also necessary to find the albumen itself.

Another significant example of erroneous conclusion is the discussion of Kriss (1959) concerning the origin of free nitrogen in waters of the Black Sea by denitrification.

Kriss has written: "Such sharply defined activity of denitrifying microorganisms in Black Sea muds and the great numbers of these organisms attest to a high intensity of nitrate reduction on the floor of the sea. It is clear that processes take place in the Black Sea muds that lead to the formation of nitrates and nitrites in such quantities that a level of denitrifying phenomena is attained indistinguishable from that observed in soils" (pp. 231-232).

The logical course of reasoning reduces to the following. There is free nitrogen in the deep waters of the Black Sea, and analyses disclose the presence of denitrifying bacteria. Since there are no nitrates, nitrification should occur, but, immediately on forming, the nitrates decompose to free nitrogen.

The author thus comes to the erroneous conclusion that nitrification is possible in an anaerobic environment despite the data of ZoBell (1945) that the lowest potential at which this process is possible is $rH_2 = 25$.

It is precisely in this way that denitrification is frequently considered to occur in formational waters because denitrifying bacteria are present, even though nitrates are generally absent in the formational waters. It is necessary to keep in mind that the reduction of nitrates may be due to a number of reductases and may not be connected with the presence of any specific ferment.

Geologists, in speaking of the activity of microorganisms, generally fall into one of two extremes: either they deny the significance of such activity entirely, as Strakhov (1947, 1948) has mostly done, or they ascribe all the uncertain genetic processes associated with oil, hydrogen sulfide and other substances to microorganisms.

There is thus a widespread opinion among geologists that methane may be the source of organic material or of energy for Vibrio desulfuricans in the formation of hydrogen sulfide in formational waters at oil fields (Sokolov, 1956). However, special studies on the physiology of this organism, made by Sorokin (1957), have shown that desulfurizing bacteria cannot use methane as their energy source.

A new method of appraising the activity of microorganisms in diagenesis is based on an analysis of ratios of stable isotopes in the initial material and in the final products formed by the given process. The lower the mass of a stable isotope, of sulfur let us say, the more easily it participates in the reduction of sulfates. Thus, the ratio S^{32}/S^{34} in hydrogen sulfide that formed by the action of desulfurizing bacteria will be greater than in the initial sulfates. In abiotic processes the ratios will be the same. This question is discussed in the work of Thode, Wanies, and Wallouch (1954) on the origin of sulfur in the salt domes of Louisiana.

In some circumstances such questions may be solved by chemical analysis; an example is the question concerning the role of microorganisms in the anaerobic decomposition of oil. In analyzing the ratio of argon to nitrogen in petroleum gases, one may conclude, when argon is absent, that the nitrogen formed by decomposition of the soil and did not come from the air. This makes it possible to determine if oil has been decomposed, with the formation of gaseous fuels.

Lastly, questions concerning the rate of individual processes in the natural cycle of a substance may be solved by using methods employing radioactive isotopes, directly in nature or by setting up experiments with isolated samples of water under conditions most closely approximating those in nature.

But the model experiments must be set up under conditions most nearly approaching natural conditions. On the basis of laboratory data one cannot draw conclusions concerning the intensity of natural processes. Let us cite two examples.

Butlin and Postgate (1952) investigated four lakes in the vicinity of Cyrenaica in North Africa. The lakes are supplied hydrogen-sulfide water from springs, and the content of molecular sulfur in the muds amounts to as much as 50% of the dry weight.

Microbiological investigations have shown that in water, and especially in muds, there are great numbers of desulfurizing and colored sulfur bacteria. The latter form a red carpet that covers shallow-water segments of the lakes.

The authors concluded that the sulfur accumulation was due to the joint action of these groups of bacteria.

Laboratory experiments, in which desulfurizing bacteria have been grown on a medium with the addition of 0.1% sodium malate together with cultures of Chromatium sp. and Chlorobium sp., completely support the conclusions of the authors. Molecular sulfur was deposited on the floors and the walls of the vessels.

Thus, Butlin and Postgate considered the activity of desulfurizing bacteria to be the principal source of hydrogen sulfide in the lakes, believing the photosynthetic products of the colored sulfur bacteria to be the organic material, since there were no other sources of organic material in the lakes.

But simple computations (Ivanov, 1957a), based on knowledge of physiology of these organisms, have shown that the organic material in the given lakes is insufficient to permit the formation of the amount of hydrogen sulfide present by the reduction of sulfates. It is clear that the principal quantity of hydrogen sulfide comes from underground springs, which the indicated authors mention.

Murzaev (1937; Murzaev, 1950) came to the same erroneous conclusions in transferring laboratory data to actual aqueous basins. The author observed the intensity of sulfate reduction on Van Delden medium, reaching 1930 mg H_2S/liter for 19 days. These results he transferred to a lake 3 km in diameter with a 10-m hydrogen-sulfide zone, and he computed that for 100 days in the summer period 407 tons of sulfur or 45 kg/m^2 should form on the bottom of the lake.

In this case A. S. Uklonskii (1940) was correct when he wrote: "On the basis of the Van Delden experiment (having nothing in common with natural conditions, since asparagine, sodium lactate, and other constituents were added), an extremely bold conclusion is drawn concerning the quantity of sulfur deposited on the floor of the basin" (p. 231).

In order that the geological activity of microorganisms may be evaluated, the following analyses should be made.

1. Quantitative or, at least, qualitative analysis of the given group of microorganisms.

2. Chemical analysis of the water or of samples collected with due consideration to the ingredients whose alteration is assumed to result from the action of the microorganisms.

3. Study of the extent to which the ecological conditions (salt concentration, rH_2, O_2, H_2S, biophores, etc.) are favorable for the development of the given group of microorganisms.

4. Exhaustive collection of data on the geology and hydrogeology of the deposit where the investigated process is at work.

5. Acquisition of exhaustive information on the physiology of the given group of organisms and a study of the effect of specific factors on the growth and life activity of these organisms, factors that the organisms encounter in nature.

6. The setting up of experiments most nearly approaching natural conditions, particularly with the employment of the highly sensitive method of radioactive isotopes.

7. Analysis of the ratios of stable isotopes in the initial material and in the final products of substances subjected to the assumed geological activity of the microorganisms.

A clear formulation of the question of investigation and a detailed treatment of the materials for following the above plan will permit one to draw the proper conclusion concerning the geological activity of microorganisms.

The papers of Ivanov (1957a, 1957b, and 1958) may be pointed out as examples of the proper approach to the evaluation of the geological activity of microorganisms; these have to do with studies of the origin of sulfur in the deposit of Shor-Su.

A detailed analysis of the geologic environment showed that the deposit of sulfur rests on the same beds as an oil deposit. Employment of isotope methods made it possible to ascertain the biological nature and intensity of hydrogen sulfide generation in the formational waters of the oil deposit.

A hydrogeological study made it possible to discover that hydrogen-sulfide waters advanced upward along the bed in the sulfur deposit and that no fresh water was received from the Gessan Range.

A determination of the oxidation-reduction potential and chemical and bacteriological analyses made it possible to find out that deep waters and formational waters become mixed in the environs of the sulfur deposit and that conditions are favorable for the present-day formation of sulfur through the action of Thiobacillus thioparus.

The use of the isotope method made it possible to ascertain what part of the sulfides is oxidized purely by some chemical process.

These investigations have been discussed in detail in the paper by Ivanov (present volume).

In conclusion it should be noted that a study of the geological activity of microorganisms in nature, a recognition of the systematic pattern of the processes, should make it possible for us to control these processes.

Conslusions

1. By geological activity of microorganisms we understand the participation of these organisms in processes of alteration or redistribution of substances and elements in nature.

2. We must not, from the number of microorganisms alone, pass judgment on the occurrence of any biological process. It is necessary to keep in mind the ecological conditions and the presence of an altered substratum.

3. A number of examples are pointed out, showing what methods of investigation must be used in solving actual problems concerning the geological activity of microorganisms.

LITERATURE CITED

Bruevich, S. V. 1953. "The chemical and biological productivity of the Black Sea," Trudy Inst. okeanologii, 7, No. 11.

Butlin, K. R. and Postgate, J. R. 1952. "The microbiological formation of sulfur in Cyrenaican Lakes." Biology of Deserts. Inst. of Biol., London, p. 12.

Isachenko, B. L. 1958. "The origin of sulfur deposits," Trudy Inst. mikrobiol. AN SSSR, No. 5.

Ivanov, M. V. 1957a. Dissertation: The Role of Microorganisms in the Formation and Destruction of Deposits of Native Sulfur [in Russian], OBN AN SSSR.

Ivanov, M. V. 1957b. "Participation of microorganisms in the formation of sulfur deposits in Shor-Su," Mikrobiologiya, 26, No. 5.

Ivanov, M. V. 1958. "The use of isotopes in studying the role of microorganisms in the formation of the Shor-Su sulfur deposit," Trudy Vses. n.-tekhn. konf. po primeneniyu radioaktivnykh i stabil'nykh izotopov. In the No.: Study of Animal Organism, the Fishing Economy, and the Food Industry [in Russian].

Kriss, A. E. 1959. Marine Microbiology [in Russian], Izd. AN SSSR.

Kriss, A. E. and Rukina, E. A. 1949. "The origin of hydrogen sulfide in the Black Sea," Mikrobiologiya, 18, No. 4.

Murzaev, P. M. 1937. Genesis of some sulfur deposits of the USSR. Econom. Geol., 32, No. 1, p. 69.

Murzaev, P. M. 1950. "Possible methods of accelerating natural processes of forming and accumulating natural sulfur," Doklady Akad. Nauk SSSR, 72, No. 2.

Sokolov, V. A. 1956. The Migration of Gas and Oil [in Russian], Izd. AN SSSR.

Sorokin, Yu. I. 1957. "Concerning the capacity of sulfate-reducing bacteria to use methane for reducing sulfates to the hydrogen sulfide," Doklady Akad. Nauk SSSR, 115, No. 4.

Strakhov, N. M. 1948. "Iron-ore facies and their analogs in earth history," Trudy Inst. geol. nauk AN SSSR, 73, geol. seriya, No. 22.

Strakhov, N. M. 1948. "The true role of bacteria in the formation of carbonate rocks," Izv. AN SSSR, seriya geol., No. 3.

Thode, H. G., Wanies, R. K., and Wallouch. 1954. "The origin of native sulfur deposits from isotope fractionation studies," Geochim. et cosmochim. acta, 5, No. 6, p. 286.

Uklonskii, A. S. 1940. The Paragenesis of Sulfur and Petroleum [in Russian], Tashkent.

Zo Bell, K. 1945. Marine Microbiology, Baltimore.

THE GEOLOGICAL ACTIVITY OF BACTERIA
AND ITS EFFECT ON GEOCHEMICAL PROCESSES

M. A. Messineva

(All-Union Scientific-Research Geological-Prospecting Petroleum Institute, Moscow)

The surface shells of the earth are characterized by a great variety in composition, phase state, and thermodynamic characteristics of the substances composing them.

The envelope of "activity of living substance," the biosphere of the earth, includes the troposphere, the hydrosphere, the stratisphere (the part of the lithosphere consisting of stratified rocks), and also some segments of the shell of metamorphic and granitic rocks.

In the biosphere, as in the entire crust of the earth, a state of unstable equilibrium is characteristic. Changes occur continuously in composition, properties, structure, and energy content of this envelope; these changes may be cyclical, sometimes are accompanied by irreversible reactions.

The dynamics of geochemical processes is determined by many natural factors: temperature, pressure, physicochemical conditions, presence of radioactive elements, and the effect of living substances.

The effect of living substances on geochemical processes is manifested on a tremendous scale and may appear in quite different forms. These forms include the accumulation of energy during photosynthesis, the accumulation of living organisms in great masses, and the concentration of organic and mineral substances in the remains of dead organisms.

Microorganisms have a special and very substantial role in geochemical processes. As the remains of plant and animal life break down, bacteria produce synthesis of new complex organic substances in the shells of the earth that contain no other organisms and that are characterized chiefly by the destruction of complex high-molecular compounds. As a result of bacterial action, substances may accumulate that not only have more complex structures, but may also contain more energy.

According to Vernadskii (1955), the geological activity of bacteria (in the broad sense) should signify all the processes effected by bacteria or occurring as a consequence of the development of bacteria in one of the earth's shells: in the atmosphere; in continental, marine, or underground waters; in soils; in rocks; etc.; since all these processes necessarily leave their imprints on the further course of substance and energy transformations with time.

At the present time the geologic activity proper of bacteria is understood in a narrower sense and is considered to relate chiefly to changes in the sedimentary shell of the earth, in ground water, in deep gases, and in mineral deposits.

The possible geological activity of bacteria and the effect of these organisms on geochemical processes were first discussed in papers of Russian scientists as early as the end of the last century.

In 1897 Vinogradskii advanced the idea that microorganisms played a dominant role in the cycle of substances in nature, and this idea gained support and further development in the classic works of V. L. Omelyanskii, B. L. Isachenko, and many of their students (Isachenko, 1927).

The works of V. O. Tauson (1936, 1947, 1948, 1950) have been of great importance in guiding the studies of geological activity of bacteria. It is in these papers that one may find the first comparisons of the physiology of various bacteria with the possible geochemical activity of the bacteria in different geologic epochs. The presently existing extensive literature on petroleum microbiology is cited in the monographs of ZoBell (1945) and Beerstecher (1954).

The Origin of Microorganisms in the Biosphere

It has now been ascertained that the interior of the earth is not sterile. Deep-lying water, many sedimentary rocks, and even porous volcanic rocks contain bacteria. In thin sections and mounted preparations of ancient rocks, pre-Paleozoic and Paleozoic, bodies of bacteria have been found that still preserve the capacity to stain with erythrosin and other dyes. In some rocks, when the material is placed on a proper culture media, viable bacteria appear.

In analyzing the surface envelopes of the earth, it is possible to discover probable courses of origin of deep-occurring microflora (Fig. 1).

As may be seen from this scheme, the principal and systematic course of development of microflora in sedimentary strata depends on interaction between the lower layers and the surface envelopes. In this process important factors are the compositions of the corresponding envelopes or shells (the habitat of the bacteria), the number of bacteria, and the rates at which the bacteria multiply and die. It is known that the boundaries of the gas-soil, gas-water, and water-sediment phases are literally "saturated" with bacteria. The biomass of bacteria in the organic material of soil is expressed by integral percentages. The biomass of bacterial plankton and periphyton (organic incrustation or scum) is commensurate with the biomass of animals and plants. The biomass of bacteria in the sediments of continental water constitutes from a few percent to 15% of the organic material. This statement refers not only to lacustrine basins (Kuznetsov, 1952), but also to the sandy bottoms of rivers where the total content of organic material is not large but the mass of bacteria is one of the principal components (Messineva, 1960). The biomass of bacteria in the deposits of salt lakes, estuaries, and seas is expressed in kilograms per cubic meter (Remezova, 1950; Messineva, 1955; Messineva and others, 1955).

The rate at which bacteria multiply in natural conditions is apparently much lower than the rate on culture media. Direct observations of the multiplication rate of bacteria in fresh-water sapropel were made by S. I. Kuznetsov (1952). It was found that the bacterial cell divides once every two weeks.

The death rate of bacteria under natural conditions is obviously the same as the rate of multiplication since the number of bacteria in a particular biomass remains constant and is the most characteristic feature of a given biotope (soil, water, different types of sediments, etc.). The death of a bacterial cell is generally associated with autolysis: with the destruction of coordinated ferment action at a definite stage of development of the cell and with self-digestion by action of the ferments proper. In this the cell disappears as a particle and "traces of its presence" may be detected only in the manifestation of fermentative activity in the soil or rock (some information on fermentative activity in rock will be given below). The action of ferments after the death of bacterial cells, and also the action of ferments on animals and plants in buried sediments, leads to hydrolytic decay of the complex molecules, to partial destruction, and, probably, to disintegration of the harmful exchange products of the bacteria. The direct consequence of autolytic fermentative processes is the preparation of a substratum for new propagation of bacteria, for a new "outburst of life" in the sediment already covered by later deposits. It should be noted that this described phenomenon is not observed on all sedimentary rocks, but only in those that have been described by the author as "geochemically active."

In "geochemically inactive" sediments and rocks the main bulk of buried bacteria is apparently fossilized. The process of fossilization may be very rapid when electrolytes are present in sufficient quantities; this is especially true for adsorbed cells. In this process part of the albuminous substance must be preserved in the fossilized cells, and this determines the capacity of the cells to take stains. It should be stated that most of the bacteria observed in preparations and thin sections are represented by fossilized cells. Their presence and their number are of interest in the study of paleontology and paleofacies. For an objective evaluation of the geologic activity of bacteria, the results of determining the quantity of bacteria must be compared with the occurrence of viable forms, with direct analysis of the intensities of the bacterial processes.

When sediments are covered by younger deposits, diagenetic processes associated with redistribution of the more mobile phases of the sediments (gases, aqueous solutions) may cause redistribution of the bacteria along the bed, and may also cause local concentrations of bacteria. Such phenomena have not yet been sufficiently studied. However, there are facts that point to the redistribution of bacteria during congelation of colloidal systems of sediment, during the formation of nodules, concretions, and microscopic zones. Local zones of bacterial accumulation also form during burial of large animal fragments or during the accumulation of a homogeneous mass of animal and plant remains.

There are data on the occurrence of filterable types of bacteria in some sediments. Particles 0.2 μ across that are clearly stained by erythrosin are found in imprint preparations and in thin sections. The viability of filterable forms is confirmed by the formation of secondary cultures when the filtrates are planted directly on culture media. The significance of these forms is not yet clear.

In transferring material (for culture) from sedimentary rocks brought up from various depths, it has been noted (as a rule) that there is a predominance of sporal forms; these sometimes constitute up to 100% of the viable bacteria. A similar distribution of soil microflora, noted by Mishustin (1953), displaying a predominance of spore-bearing bacteria in sedimentary rocks, serves as an indicator of the absence of easily hydrolyzed substances. It should be emphasized that in order to obtain the percentage content of spore-bearing bacteria in rock it is most advisable

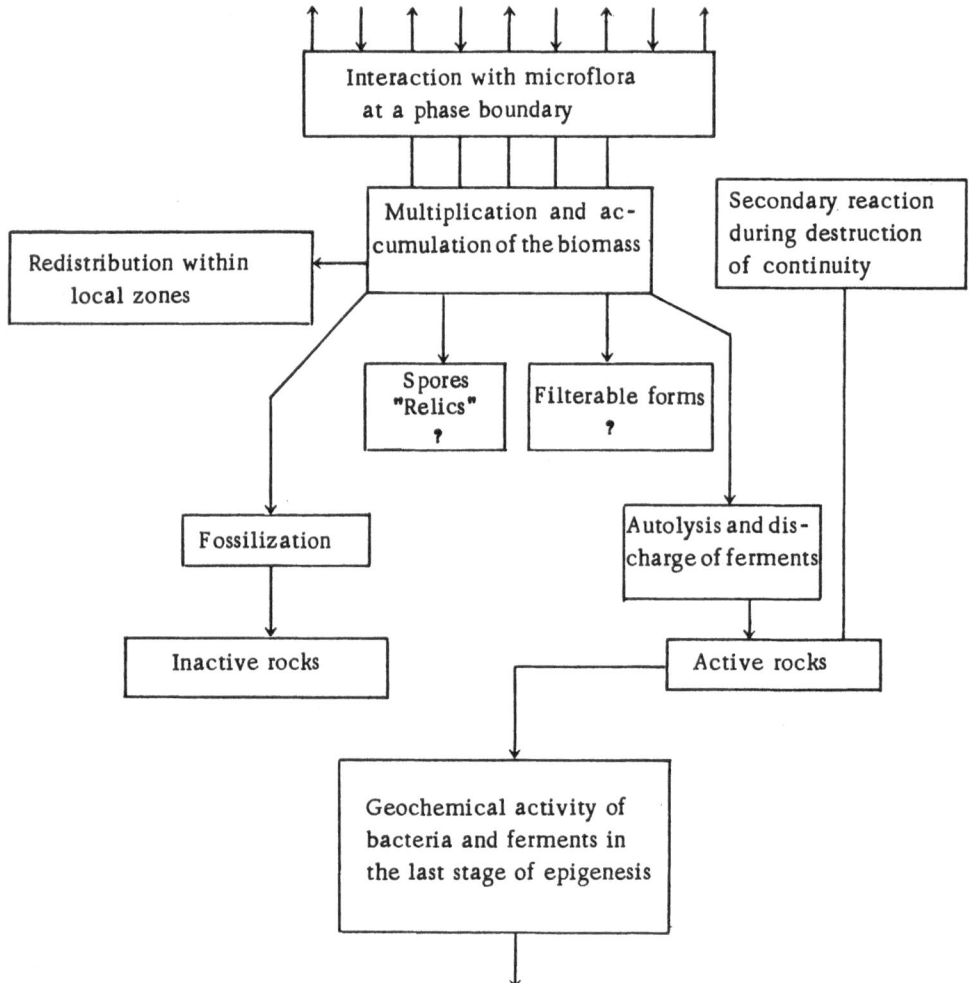

Fig. 1. Scheme of formation of bacterial shells of the biosphere.

to define the frequency of taking samples, since the drilling muds used when boring holes have different ratios of sporal and nonsporal bacteria, generally with a predominance of the latter.

Bacterial spores are also seen during microscopic examination of sediments. The development of some bacterial relicts may be associated with the formation of spores.

One cannot deny the possibility of "secondary colonization" inside the earth by penetration of microorganisms through available ruptures in the continuous sedimentary cover. Powerful streams of gas, discharges of water, movements of fluids under great pressure from below upward undoubtedly produce wall effects, the development of cavities, and these make it possible for bacteria to move downward in the countercurrent.

Tectogenesis and the associated disruption of continuity, the appearance of fractures on the limbs of structures, the formation of local vacuum cavities, the secondary penetration of fluids during migration—all these are possible means of movement and penetration of bacteria into the interior of the earth.

The General Pattern of Distribution of Microorganisms in the Various Shells of the Earth

The distribution of microorganisms in the present-day surface shells of the earth is very irregular. Within the two principal geospheres—the geosphere of land and the geosphere of the sea—the number of microorganisms and the specific variety are determined by the different ecological conditions, and also by the character of the biocoenotic ratios of the different organisms constituting the natural biotopes. Many specific peculiarities are also observed in the distribution of microorganisms in water and sediments of stagnant and through-flowing continental basins.

The difference between the number of bacterial cells seen under the microscope and the number of living bacteria determined by growths on culture media is generally very large. Even from present-day soils, growths on various nutrient media reveal no more than 3% of the observed bacterial cells. A similar relationship is also observed in fresh-water and marine sediments: a great discrepancy is noted between the total number of bacteria and the number of bacteria growing on culture media. One of the principal causes of this discrepancy is probably the adsorption of bacterial cells on the surfaces of colloidal particles of the substratum.

Data from numerous investigations have shown that the adsorption of bacteria is directly dependent on the quantity of the colloidal fraction in the investigated sediment or soil. This same relationship is also characteristic for the intensity of bacterial processes in nature. The greatest intensity of the bacterial processes in sandy stream bottoms, in which extremely rapid mineralization of organic material favors self-purification. In clay and clay-silt sediments the processes of mineralization are slowed down, despite the fact that the number of bacteria in the clay sediments is considerably greater than in sand-silt sediments. As a consequence of this retarded mineralization in clays there is a gradual accumulation of organic material. In sapropelic sediments, which have a high content of organic colloids, the rate of mineralization of organic substances is also low, although the number of bacteria in such sediments amounts to a billion cells per gram of sapropel (dry weight). Direct observations have shown that when sapropel is mixed with sand the intensity of the microbiological processes is increased many times (Messineva and Gorbunova, 1940, 1946). As our further investigations have shown, the grain-size distribution in rocks has a substantial effect even on such important factors as the use of moisture by bacteria. It becomes obvious that it is necessary above all to take into account the peculiarities of a biotope or of the facies of a deposit in order to evaluate geological activity of bacteria.

Frequently attempts are made to infer a relationship between distribution of bacteria and the quantity of organic material and biogenic elements. There undoubtedly is such a relationship, but it is true within identical or comparable biotopes or facies of sediments.

The nature of a biotope also determines the thickness of the "bacterial film" (mass accumulation of bacteria), which forms at the boundary of two phases. In continental waters the thickness of the layer of massive development of bacteria at the water-sediment boundary depends on the development of benthonic forms of plants and animals. The same applies to marine sediments of the littoral zone. At great depths bacteria are observed to accumulate in a very thin layer one to a few millimeters thick.

The irregular distribution of bacteria is the most characteristic feature of the present-day surface shells of the biosphere. In past geologic epochs, as at the present time, the earth had zones of bacterial concentration, zones of limited distribution, and zones where bacteria were almost completely absent. It is probable that among microorganisms there are also ephemeral forms, able to persist through unfavorable conditions (such as rapid desiccation of sediments) and to give a new short-lived "burst of life" during changes in these conditions.

The geometry of living material is still more complex than the geometry of the geospheres and the shells of the earth. There are therefore no grounds for restricting the manifestation of geological activity of bacteria to rocks of any particular age or specific depth of occurrence. It is very likely that at certain zones of the earth's crust the boundaries of the biosphere are found below proterozoic rocks. The actual depths to the boundaries of the biosphere certainly vary in different parts of the world, since the geothermal gradient and the associated rise in temperature change according to the geologic structure and the tectonics of the region.

In geochemistry it is necessary to examine a bacterial cell not only as a living or viable unit, but also as a particle of definite size, possessing all the properties of colloidal particles and having a definite charge. The physicochemical peculiarities of a bacterial cell favor the attachment of the cell, in a layer of water, to simple agglomerations consisting of organo-mineral complexes and the living substance of microorganisms.

Of especial significance in geochemical processes is the capacity of microorganisms to give off ferments to the external environment both as manifestation of life activity and through autolysis after death. The conditions necessary for preservation of the activity of ferments and for producing fermentative catalysis are less confined than the conditions necessary for the life activity of the microorganisms themselves. As already suggested, the development of bacteria in many situations may be limited by inadequate access to organic material and biogenic elements, by inadequate moisture, by the presence of exchange products, and by other factors. These restrictions have no significance for ferments. The subsequent effect of ferments, therefore, after autolysis of the bacterial cells themselves, may obtain for a long period of time.

The view concerning possible fermentative activity in soils and rocks has been advanced by Omelyanskii, Gubkin, and others. The first stage of biogeochemical transformation may be clearly distinguished; in this stage the characteristic features are hydrolytic decomposition of complex high-molecular organic compounds and accumulation of a biomass of bacteria. Beyond this, the body of bacteria should be considered as a newly formed organic substance of different composition, possessing different properties from those of the original substance. As the bacterial biomass accumulates the colloidal fraction in the sediments increases and ion exchange is retarded in the aqueous phase of the sediment. The significance of bacteria in diagenesis of sediments has received proper attention in the work of Strakhov (1956), who considers the so-called biogenic factor a fundamental and leading factor in the diagenetic stage.

It should be noted that one still encounters in some microbiological papers the erroneous, restricted views that the diagenesis of sediments is only a stage of dehydration, the stage of "squeezing water out," as a result of which the participation of bacteria in diagenesis is thought to be very limited.

A study of the actual phenomena in nature has shown that during diagenesis there occurs not only dehydration but increased mineral formation, separation of a gaseous phase, change in the degree of hydrophilic and hydrophobic properties of the colloids, increased heterogeneity of the sediments, and change in structure of the sediments. Microorganisms play an important role in all these phenomena, their effect being manifested in the accumulation of the mass of colloids, in the separation of gases, and in a different type of chemical and physicochemical transformation. Lastly, the separation of ferments during autolysis of bacteria makes possible transformations to the succeeding stages of lithogenesis.

The structural development of the sediment and the increase in heterogeneity of the soil are especially important in the subsequent synthesis of complex authigenic compounds. It is well known that the action of ferments is reversible. Hydrolytic processes take place in an excess of moisture in a hydrophylic environment; synthesis occurs in a hydrophobic environment with the elimination of water. The synthesis of complex substances, such as hydrocarbons, humic acids, and others, is thus effected in a hydrophobic environment with a limited amount of water. In this connection there is special significance in studying the natural conditions in which a slight amount of moisture in sediments does not impede the development of bacteria. The general views of microbiologists that the content of water is the basic factor regulating the development of bacteria do not correspond with the facts; such views require only consideration of the liquid nutrient media during water excess, an attitude that completely distorts the systematic pattern observed in nature (Messineva and others, 1955; Messineva, 1955).

In studying the activity of bacteria we have employed the following method for a number of papers. The rocks are ground aseptically (in a mortar as described by V. O. Tauson), placed in a sterile flask, and the substance whose transformation we wish to ascertain is added to them in sufficient quantity necessary for analysis, but not in amounts exceeding 3-5% of the weight of the rock. Some of the experiments are conducted with the moisture content found in nature, others with moisture added. In these experiments the bacteria are not grown on culture media, but direc ly in rocks, i.e., under conditions approximating those in nature. During the last 10 years more than 700 of such experiments have been conducted, using rocks taken from drill holes at depths down to 500 m (Maikop), and also using a number of present-day marine and fresh-water sediments. In these studies it was discovered that all the investigated sediments and rocks could be divided into two basic types.

1. Heterogeneous sediments and rocks (containing organic material, nitrogen, and phosphorus), in which the number of bacteria, however slowly, increased without the addition of moisture. In these rocks the addition of moisture leads to rapid reworking of the introduced material. Such rocks may be tentatively designated as geochemically active.

2. Homogeneous, well-sorted clays, as a rule, not allowing the development of bacteria even after addition of moisture. In some experiments a decrease in number of bacteria was noted after the investigations. In such experiments, when rocks of this second type proved to be barren even of fermentative properties (special supplementary analyses), the rocks were tentatively designated geochemically inactive (Messineva, 1955).

No connection has been found between the number of bacterial cells detected in a sample and the geochemical activity of the sample. From this we may draw the conclusion that an appraisal of the participation of bacteria in geochemical processes by the number of cells identified by direct count or by restricted growth on selected media is improper, and all conclusions by authors maintaining this should be re-examined.

The differences in geochemical activity in rocks are not determined by depth of occurrence or by age of the rock; they are functions of lithologic-geochemical and facies features. The conclusions we have arrived at once more give support to the validity of V. I. Vernadskii's judgment concerning the discontinuity and complex geometry of the various shells of the earth, especially the biosphere.

The Distribution of Bacteria along the Vertical Section

The distribution of bacteria in a vertical section of the earth is as irregular as the surface distribution. The distribution is determined by changes in facies environments throughout geologic history in each segment of the earth's crust accessible to study. The probable limit for the existence of bacteria and for the activity of ferments is the zone of temperatures above 100°, which is found at various depths in different regions, in places being more than 4-5 km below the surface.

Relevant studies by the author have shown that bacterial activity (especially in rocks not saturated with water) is more frequently discontinuous along the vertical than is the geochemical activity of ferments. This fact is explained by special conditions required for the life activity of bacteria that are not required for the activity of ferments.

The study of bacterial distribution along the vertical section has been made by the author not only for a general systematic arrangement of the facts, but also because of the hypothesis evinced by geologists that there is a so-called attenuation curve for the activity of bacteria. This hypothesis that the activity of bacteria is restricted in proportion to the depth of occurrence of the rock had dire consequences for geochemistry, since for many years it served geologists as an argument to demonstrate the inorganic origin of oil (Kudryavtsev, 1955). On closer examination of the hypothesis of this "attenuation curve," not only is it found that facts do not support it, but it has been proved incorrect methodologically (Messineva and others, 1955).

A marked decrease in number of bacterial cells is observed only at a depth of one meter (sometimes several meters) in sediments, i.e., within the first bacterial shell at phase boundaries (Fig. 2). This decrease in number of cells is by no means always accompanied by lower geochemical activity of the bacteria. Furthermore, beyond the upper shell, bacterial distribution, as the investigations have shown, does not depend on the depth of occurrence of the rocks.

Authigenic Bituminous Material and Humic Acids Due to the Work of Bacteria and Ferments

The possibility of newly formed complex high-molecular compounds in sedimentary rocks is generally demonstrated by comparing chemical compositions of rocks in the section. An increased content of any particular component in the older rocks as compared with overlying rocks (in identical facies) is proof that transformation has occurred. Direct experiments are considered impossible by most geologists and geochemists because of the "nonreproducibility" of natural conditions, especially of geologic time. However, experiments set up by the author on various rocks have demonstrated the possibility of considering results for periods of time of one or two years.

It has been ascertained that authigenic bituminous material and humic acids form through the action of bacteria and ferments. In some samples bacteria were found to have the capacity to destroy bituminous material and to increase the content of insoluble organic compounds in the rock.

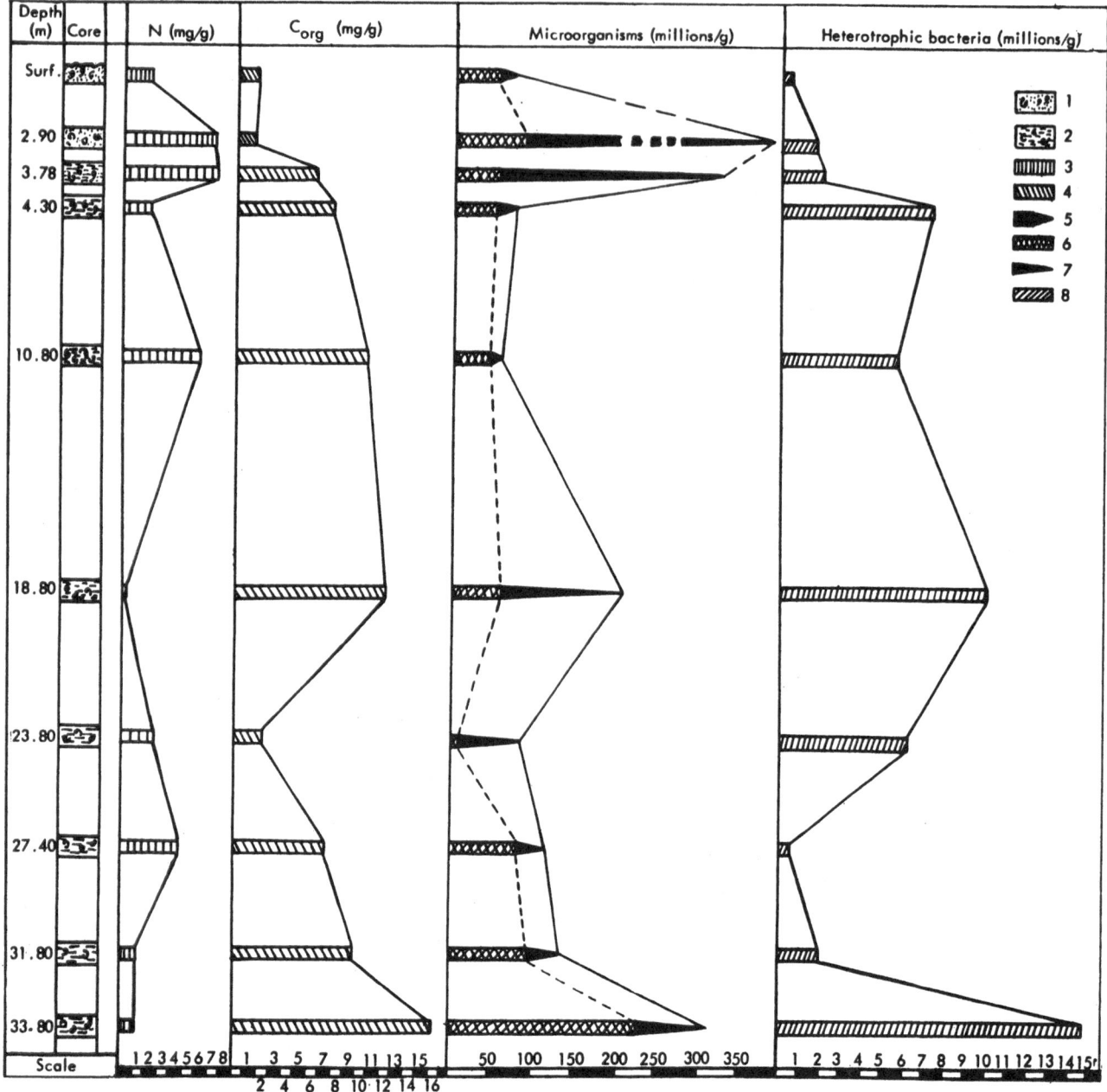

Fig. 2. Distribution of microorganisms and content of organic material in sediments of the Sea of Azov (from the section of a drill hole). 1) Surficial sandy sediments with molluscan fauna; 2) clay with molluscan fauna; 3) total content of nitrogen (milligram per gram of dry sediment); 4) organic carbon (milligram per gram of dry sediment); 5) total number of microorganisms per gram of dry sediment (millions of cells); 6) cocci; 7) bacilli; 8) heterotrophic bacteria (millions per gram of dry sediment).

In experiments on converting vegetable oil, the fermentative processes diminished the fraction of material soluble in chloroform and increased the fraction of insoluble organic material. Living organisms converted a considerably larger amount of organic material to the insoluble state; part of this material was broken down to CO_2. Thus the overall content of material soluble in chloroform was reduced to half the original amount during the experiment. One's attention is drawn to the newly formed fraction of oils and petroleum ether tars in the chloroform extract; this is of great interest for the hypothesis of the organic origin of petroleum.

The action of bacteria and ferments in clay and layers of silt leads to destruction of part of the bituminous material in these deposits. In silt authigenic bituminous material forms, particularly chloroform-soluble bitumen A, and there is an increase in content of elemental hydrogen in this bitumen. In the tested samples of these two rocks there was also a difference in the fate of the humic acids: in the clay the amount of humic acids increased 30%, but in the silt layer it decreased 40% in comparison with the control.

Experiments on the study of changes in content of bituminous material and of humic acids during decaying action of the brown alga Cystosira barbata have shown that the organic material is transformed by the action of epiphytic bacteria found on the surfaces of the algae and also by the action of ferments of the algae and bacteria. The oxidation-reduction potential during the experiments did not rise above −230 mv. The given experiments demonstrated the authigenic formation of humic acids, and also showed changes in the quantity of bituminous components. The newly formed humic acids in these experiments developed in an anaerobic environment.

Conclusions

1. Microorganisms are among the important agents of the various geochemical processes.

The influence of bacteria is manifested in three basic directions: a) the consequence of propagation and exchange of material of living vegetative forms; b) processes associated with the dying of bacteria, their autolysis, and subsequent protracted action of ferments; c) accumulation of a biomass of bacteria, characterized by authigenic complex organic substances and by changes in the physicochemical state of the sediments and in the structure of sediments and in the structure of sedimentary rocks.

2. The distribution and activity of microorganisms in the biosphere are defined by the principal parameters of the biotope or of the facies of the deposits.

The geochemical activity of bacteria is manifested irregularly both areally and vertically in the earth.

The manifestations of geological activity and geochemical activity of bacteria are subject to the general laws of geometry of the biosphere and of geometry of the earth shells as formulated by Vernadskii.

3. Sedimentary rocks may be divided, regardless of age and depth of occurrence, into "geochemically active" and "geochemically inactive" rocks, i.e., in the latter type, rocks in which there is no actual manifestation of bacterial or fermentative action.

4. In "geochemically active" rocks, as shown by experiments on the rock material itself, authigenic humic acids and bituminous substances are formed, and in the latter there appears a newly formed oil fraction.

5. Further study of the geological activity of bacteria should be accompanied by the development of new methods, both for laboratory and for field work. The new methods should guarantee the possibility of observing the action of bacteria not on artificial nutrient media, but under natural conditions, or under conditions that approach the natural.

6. The following meagerly treated questions deserve special attention: a) the dynamics of bacterial propagation under natural conditions; b) the processes of dying in bacteria, especially the processes of fossilization and autolysis; c) an explanation of the relationship between bacteria and other particles in the colloidal fraction, a study of the mechanism of adsorption and the coagulation of organo-mineralic complexes.

LITERATURE CITED

Beerstecher, E. 1954. Petroleum Microbiology, Van Nostrand. Princeton.

Isachenko, B. L. 1927. "Microbiological investigations of muddy lakes," Trudy Geol. kom., novaya seriya, No. 148.

Kudryavtsev, N. I. 1955. "The current status of the problems concerning the origin of oil," in the Collection: Materials on the Discussion of the Origin and Migration of Oil [in Russian], Kiev.

Kuznetsov, S. I. 1952. The Role of Microorganisms in the Cycle of Substances in Lakes [in Russian], Izd. AN SSSR.

Messineva, M. A. 1940. "The fermentative properties of fresh-water muds," Byull. MOIP, otd. biol., 12.

Messineva, M. A. 1955. "Experimental proofs of the possibility of forming oil from organic material (appearance in rotten material)," in the Collection: Biocoenosis of Overgrowths (Scums and Incrustations) as Bioabsorbents [in Russian], Izd. MGU.

Messineva, M. A. and Gorbunova, A. I. 1940. "The change in content of nitrogen compounds in sapropel as a result of microorganism activity," Mikrobiologiya, 9, No. 7.

Messineva, M. A. and Gorbunova, A. I. 1946. "The decomposition of microphytes in fresh-water lakes and the participation of their remains in the formation of lacustrine mud deposits," Izv. AN SSSR, seriya biol. No. 5.

Messineva, M. A., Bogomolov, V. I., and Gorskaya, A. I. 1955. "Factors controlling the conversion of organic material to oil," in the Collection: The Origin of Oil [in Russian], Gostoptekhizdat.

Mishustin, E. N. 1953. Microorganisms and the Fertility of Soil [in Russian], Trudy Konf. po voprosam pochvennoi mikrobiol., Izd. AN SSSR.

Remezova, T. S. 1950. "Microbiological characteristics of bottom deposits in the aqueous basins of the Taman Peninsula," in the Collection: Recent Correlatives of Oil-Bearing Facies [in Russian], Gostoptekhizdat.

Strakhov, N. M. 1956. "Some errors in method in the study of chemicobiological sedimentation and diagenesis," Byull. MOIP, otd. geol., 31, No. 2.

Tauson, V. O. 1936. "The evolution of microorganisms during geological ages," Arkhiv biol. nauk, 43, No. 2-3.

Tauson, V. O. 1947. The Inheritance of Microbes [in Russian], Izd. AN SSSR.

Tauson, V. O. 1948. Great Deeds of Small Substances [in Russian], Izd. AN SSSR.

Tauson, V. O. 1950. The Basic Position of Plant Bio-Energetics [in Russian], Izd. AN SSSR.

Vernadskii, V. I. 1955. Selected Papers [in Russian], 2, Izd. AN SSSR.

ZoBell, K. 1945. Marine Microbiology, Baltimore.

THE EFFECT OF ECOLOGICAL FACTORS
ON MICROORGANISMS IN OIL DEPOSITS

L. D. Shturm

(Institute of Microbiology, Academy of Sciences, USSR, Moscow)

The ecological conditions in oil deposits are complex and variable in their physical, physicochemical, geochemical, and geological characteristics.

The ecological factors are interacting, and, at present, we do not have sufficient data to form any complete picture of the total effect these factors have on the life activity of microorganisms in oil deposits.

A thorough investigation of the indicated problem should therefore precede any study of the effect of individual ecological factors. We shall pause to consider the principal factors: temperature, pressure Eh, pH, and salinity; and we shall examine briefly the value of oil as a nutrient source for microorganisms. In keeping with our appointed task, it is most suitable to illustrate the effect of these factors by examples of bacteria found in deep-lying formations of oil deposits. However, because of insufficient data, we shall consider the overall biological significance of the indicated factors in application to other materials.

According to the theory of absolute rates of chemical reactions (Johnson, Brown, and Marsland, 1942), any chemical process (oxidation, hydrolysis, synthesis, etc.) is characterized by equilibrium between an active complex and reactive substances, i.e., as applied to processes in living cells, between an active, native form of enzyme and an inactive, denatured form.

Keeping in mind the probability of denaturation, one might expect temperature to have a favorable effect on the development of microorganisms, if the temperature does not exceed some specified optimum that is characteristic of a given species or strain. For the great majority of microorganisms, constituting the extensive group of mesophylic forms, this will be a moderate temperature, on the order of 20-40°. At temperatures not exceeding the optimum, sulfate-reducing bacteria are found to be very active. The content of hydrogen sulfide in cultures of Vibrio aestuarii at 35° reaches 2308 mg/liter (Pel'sh, 1937); in cultures of Desulphovibrio desulfuricans the content is 2000-2500 mg/liter (Butlin and Postgate, 1935), and nearly 3586 mg/liter on media with salts of heavy metals that form insoluble sulfides (Miller, 1950). When the temperature rises above the maximum, the life activity of bacteria ceases and sulfates are no longer reduced. Attempts to accustom mesophylic strains to higher temperatures have not given positive results (ZoBell, 1958; Postage, 1959).

Thermophylic bacteria are much more resistant to thermal activity. The optimum temperatures for these forms range from 45 to 60°. Sulfate-reducing bacteria from sulfur-bearing limestones about 500 m deep have developed at 85° (ZoBell, 1958).

In high-temperature formational waters (about 60-70°) in the Terek-Dagestan region, sulfate-reducing bacteria are found growing at 60-70° (Kolesnik, 1955). In the hot-spring region of New Zealand, sulfate-reducing bacteria are found growing at 55° (Kaplan, 1956). Sulfate-reducing bacteria grow at temperatures ranging from 0 to 100° (ZoBell, 1958).

Bacteria of the genus Clostridium are very heat resistant (Basset and Macheboeuf, 1931; Prévot, Raynaud, Tatoki, 1951; and others).

Number of Colonies on Plates of Agar, Transferred from Cultures Grown in Broth for 48 hr at 30° and at Various Hydrostatic Pressures (after ZoBell and Johnson, 1949)

Culture	Initial no. of cells per ml	No. of colonies per ml after 48 hr at the indicated pressures, atm				
		1	300	400	500	600
Alkaligenes viscosus	700	160,000,000	<100	<100	<10	0
Bacillus cereus	1600	40,000,000	700	700	100	0
Bacillus circulans	20,000	11,000,000	265,000	265,000	200	20
Bacillus mesentericus	1900	21,000,000	60,000	60,000	10	0
Bacillus mycoides	2250	2,000,000	114,000	114,000	700	0
Bacillus subtilis	1600	14,000,000	42,000	42,000	30	8
Proteus vulgaris	2600	142,000,000	<100	100	<10	0
Pseudomonas fluorescens	10,000	95,000,000	21,000,000	5,000,000	810	80
Sarcina lutea	4700	36,000,000	67,000	22,000	400	12
Serratia marcescens	300	64,000,000	<100	<100	<10	0
Staphylococcus albus	4400	77,000,000	80,000,000	57,000	700	350
Staphyloccoccus aureus	800	156,000,000	174,000,000	9000	0	0
Streptococcus lactis	4000	273,000,000	149,000,000	70,000,000	160,000	180,000

At present some authors believe that thermophylic sulfate-reducing bacteria belong to the species Clostridium nigrificans and are not a variety of the species Desulfovibrio desulfuricans (Campbell, Frank, and Hall, 1957; Postgate, 1959).

There are two points of view concerning the causes of thermal stability of thermophylic bacteria. According to one of these, represented by Allen (1950), the rate of enzymatic synthetic reactions in thermophylic bacteria is greater than the rate of the destructive process.

According to the other point of view, supported by a number of experimental data (Militzer and others, 1950; Postgate, 1959; and others), thermophylic bacteria have enzymes of greater thermal stability than mesophylic forms.

Because of the high thermal tolerance of thermophyles, the bacterial processes may occur in oil-bearing horizons in formational waters with relatively high temperatures.

Experimental data of ZoBell and Johnson (1949) actually show that investigated microorganisms from the genera Bacillus, Clostridium, Escherichia, Sarcina, Mycobacterium, Pseudomonas, Staphylococcus, and Streptococcus— 20 species in all, used for experiments in the logarithmic phase of growth—propagate weakly at 20° and a pressure of 300 atm, but grow well at the same pressure and a temperature of 40°, i.e., at a temperature 10° above their optimum. In addition we may point out that not one of the cultures developed at a pressure of 600 atm and at temperatures of 20 or 30°, but four species— Bacillus mesentericus, B. subtilis, Escherichia coli, and Streptococcus lacti lactis— grew well at this pressure at a temperature of 40°. Thus, the unfavorable effect of high pressure, retarding growth at low temperatures, decreases at high temperatures. Similar data are to be found in earlier papers (Johnson and Lewin, 1956).

Available experimental data indicate differences in sensitivity to high pressures among the investigated species of microorganisms. Yeasts have proved to be more sensitive than bacteria. Almost all the species of yeast studied died at a pressure of 400 atm, and not one grew at pressures of 500 or 600 atm. Most species of bacteria grew at 400 atm, and some at 500 atm, but not one at 600 atm. Microbacteria possessed approximately the same sensitivity to pressure as other bacteria.

Data on the viability of different bacteria at 30° and at pressures of 300-600 atm are presented in the accompanying table.

From this table it may be seen that at a pressure as low as 300 the number of viable cells is greatly reduced at 400 and 500 atm the number becomes considerably smaller, and at 600 atm the pressure is sterilizing; more than half the cultures showed no growth.

The same results were obtained in similar experiments with marine barophylic, pressure-tolerant, and barophobic microorganisms. Some pressure-tolerant microorganisms change morphologically under pressure (Serratia marinorubra) and lose mobility (ZoBell and Oppenheimer, 1950).

On the basis of experimental data the author has attempted to formulate the connection between the relation of the studied cultures to pressure and the natural habitat of the organisms. Thus, species taken from material on the surface of the ocean react to increased pressure the same as soil microorganisms. Species taken from greater depths, where the pressure reaches 500 atm, thrive in the laboratory at a pressure of 600 atm and at a temperature of 30-40°. Sulfate-reducing bacteria from petroliferous saline formational water at a depth greater than 1-2 km formed 22 mg of hydrogen sulfide at atmospheric pressure in laboratory experiments lasting 18 hr, by hydrogenous sulfate-reduction at pressures of 400-600-1000 atm, up to 79 mg of hydrogen sulfide was formed. On the basis of the data obtained the authors have suggested that the mechanism by which pressure acts on the growth and viability of cells is assoc ated with the state of the equilibrium systems of biochemical reactions that are irreversibly suppressed at critical pressures and temperatures.

The barophylic properties are not lost, but are still to be observed after several years of culture growth at atmospheric pressure and a temperature of 4°. The author believes these properties are transmitted by heredity.

Thus, the reaction of an organism at high pressure is determined to a considerable degree by the actual conditions in its natural habitat.

The Effect of Osmotic Pressure

The state and living functions of animal or plant cells are determined to a high degree by the relationship of osmotic pressure within the cell and in the surrounding medium. The osmotic pressure of protoplasm is not of constant value for any given species or strain. It changes according to the medium and the stage of development of the culture.

Bacteria differ strongly among themselves in capacity to transmit extreme osmotic pressure and in adaptability to sharp changes in this pressure (Werkman and Wilson, 1951).

This view is confirmed by many investigations on sulfate-reducing bacteria (Van Delden, 1904; Ginzburg, Karagicheva, 1926; Bastin, 1926; Isachenko, 1927; Maliyants, 1933; Rubenchik and Goikherman, 1936; Khait, 1924, ZoBell, 1946a; and others).

These investigations have shown that sulfate-reducing bacteria living in soils and in fresh-water lakes grow in media containing small quantities of NaCl (less than 0.85%). On the other hand, bacteria living in salt-water basins (Desulfovibrio aestuarii) develop better in sea water or in media with an isotonic solution of NaCl. However, ZoBell (1958) has pointed out that freshly segregated cultures of Desulfovibrio aestuarii transmit hypertonic solutions (as compared with sea water) no better than hypotonic solutions.

The salt-tolerance of sulfate-reducing bacteria is highly variable, and is associated by many authors with the salinity of the medium in which the organism lives in nature (Ginzburg-Karagicheva, 1926; Rubenchik and Goikherman, 1940; Shturm, 1951). Cultured sulfate-reducing bacteria on media of various compositions have shown that, as a rule, they grow better in solutions comparable to those from which they were extracted (ZoBell, 1958).

ZoBell (1958) discovered bacteria growing in solutions with 20-30% HCl in highly mineralized formational waters at Great Salt Lake. However, halophilic sulfate-reducing bacteria with a salt optimum above 11-12% have not yet been discovered.

The maximum temperature for growth of sulfate-reducing bacteria in sea water lies between 40 and 45°.

A decrease in the salinity of an environment depresses the maximum temperature of growth of sulfate-reducing bacteria: at a dilution with sea water of 1 : 1, the temperature is depressed to 35-40°; at a dilution of 1 : 4, the depression is to 30-35°.

These bacteria thrive best in environments with the same osmotic properties found in their natural habitats (ZoBell, 1958).

In connection with the variability of salt-tolerance of sulfate-reducing bacteria, tests have been made on the adaptation of these bacteria to different concentrations of NaCl. According to Baars (1930), by successive inoculations of Desulfovibrio aestuarii in media with gradually diminishing contents of NaCl, it is possible to adapt marine species to a salt-free environment, and contrariwise, it is possible to adapt fresh-water forms to an environment with 30% NaCl. However, the author's experiments were not supported by work with cultures taken from a single cell (Rittenberg, 1941; according to ZoBell, 1958). It has been pointed out that, under the conditions of Baars's experiments, successive selection of more or less salt-resistant cells may occur, and the result may not be due to transmutations (Allen, 1950; ZoBell, 1958).

According to our data, sulfate-reducing bacteria in the highly mineralized formational waters of the Syzran oil field (sp gr, 1.079) have preserved fermentative activity at NaCl contents ranging from 0.5 to 15%, the maximum generation of H₂S occurring at 11% NaCl. Sulfate-reducing bacteria from the less mineralized waters of the Changyr-Tash field (sp gr, 1.0139) produced H₂S at concentrations of NaCl ranging from 0.5 to 10%, the maximum fermentative activity appearing in a medium with 6% NaCl (Fig. 1).

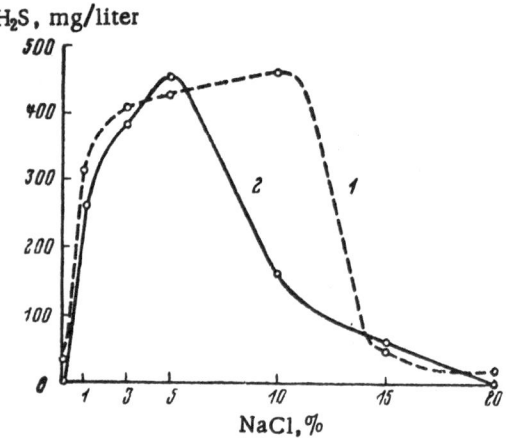

H₂S, mg/liter

NaCl, %

Fig. 1. Formation of hydrogen sulfide in a medium with varying contents of sodium chloride. 1) Vibrio desulfuricans from Lower Carboniferous formational water; 2) Vibrio desulfuricans from Tertiary formational water.

The cited data show that the osmotic properties of a medium have an effect on the fermentative activity of sulfate-reducing bacteria; this activity depends on the adaptation of the microorganisms to the salinity of the environment in nature.

Natural selection of salt-tolerant strains has been observed by Rubenchik and Goikherman (1936) in connection with seasonal fluctuations of salinity in estuaries; the optimum salt content for sulfate-reducing bacteria changes correspondingly in this process.

The Effect of pH and Eh on the Life Activity of Microorganisms

The most important physiological functions of organisms associated with enzymatic activity, the colloidal state of protoplasm, and the permeability of the cell wall are regulated and conditioned by the active concentration of hydrogen ions. The action of these free hydrogen ions depends not only on the degree of dissociation of the H-bearing electrolyte, but also on the presence of salts and on the buffering and oxidizing-reducing properties of the environment (Rabotnova, 1957).

Pel'sh (1937) believes that the effect of the pH on the reduction of sulfates is twofold: 1) a direct effect is involved in the marked predominance of H and OH ions, reflected in all living processes, and 2) an indirect effect is manifested in the average values of pH. In this interval the pH very markedly indicates the distribution of the forms of sulfur: H₂S, HS', and S". The hydrosulfide form HS' is easily transferred by bacteria; free hydrogen sulfide proves to have the most toxic effect.

Sulfate-reducing bacteria taken from soil (Starkey and Wight, 1945) grew between the pH limits of 5.5 and 8.5, but would not develop at pH values below 5 or above 9. According to ZoBell (1946b) the extreme pH limits are 4.2 and 10.5; the maximum growth has been observed at a pH of 7.

According to our data, sulfate-reducing bacteria taken from petroliferous formational waters from Lower Carboniferous strata at the Syzran field were grown in the pH interval between 5.0 and 9.60, the maximum evolution of hydrogen sulfide (700-770 mg/liter) occurring at a pH between 7.0 and 9.60. Sulfate-reducing bacteria taken from petroliferous formational waters from Tertiary strata at the Changyr-Tash field were grown in the pH interval between 6.0 and 8.8, the maximum evolution of H₂S occurring at a pH of 8.2 (about 750 mg/liter); at pH = 6.0 and pH = 9.0 no growth was observed (Fig. 2).

According to experimental data of Bass-Becking and Kaplan (1956), the relation of sulfate-reducing bacteria to pH depends on Eh under laboratory and natural conditions.

A pH of 6.2-7.9 and an Eh ranging from −50 to −150 mv are most favorable for the growth of these bacteria.

In regard to Eh, the presence of sulfate-reducing bacteria in different horizons of many structures indicates that conditions are favorable for the growth of these bacteria in these horizons. So far as we know, the actual values of pH and Eh obtaining in oil deposits will permit bacterial activity.

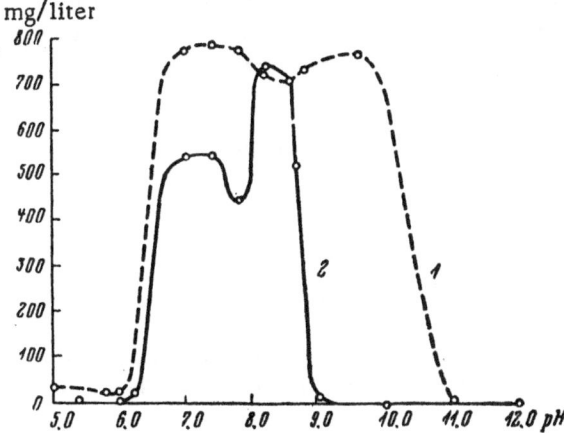

Fig. 2. Formation of hydrogen sulfide at various pH values in a medium with lactate. Symbols are the same as in Fig. 1.

The Action of Microorganisms on Oil and Hydrocharbons

Oil, the different compounds composing it, and petroleum products are easily used by microorganisms in an aerobic environment. It has been ascertained that oil may be completely consumed by microbes.

The most exhaustive survey of the numerous papers in this field may be found in Beerstecher's monograph "Petroleum Microbiology" and in the paper of ZoBell "Assimilation of Hydrocarbons by Microorganisms" (ZoBell, 1950).

Oxygen, as expected, is absent in the productive beds of sealed oil structures. Whether there are traces of oxygen in exposed deposits is unknown. Thus, the study of the action of microorganisms on oil and hydrocarbons in an anaerobic environment is of special interest in petroleum microbiology. Investigations have been made chiefly with Vibrio desulfuricans. A number of authors have noted the reduction of sulfates in a medium with oil (Uspenskii and others, 1947; ZoBell, 1947; Shturm, 1951; Kuznetsova and others, 1957; etc.).

In contrast to this, some investigators have obtained no positive results (Baars, 1930; Davis, 1956; Simakova, 1956; Updegraff and Wren, 1954). The cause of this discrepancy is not clear.

In regard to hydrocarbons, according to the investigation of Novelli and ZoBell (1944), some strains of sulfate-reducing bacteria may utilize hydrocarbons, beginning with decane and higher forms, paraffin oil, and paraffin wax.

The aromatic hydrocarbons— benzene, xylene, anthracene, and naphthalene— are not utilized, nor are the hydrocarbons of the aliphatic series, those of lower molecular weight than decane, or hydrocarbons of the naphthene series, cyclohexane. The authors present no quantitative data. Rosenfeld (1947) also notes that the high-molecular hydrocarbons of the aliphatic series are quickly decomposed by sulfate-reducing bacteria. The author has established that fatty acids are formed during the decomposition of hexadecane, but they do not accumulate in the medium; they decompose farther. The author associates the use of hydrocarbons with the presence of dehydrogenase.

According to our data, Vibrio desulfuricans strain 313 extracted from formational water from a Lower Carboniferous horizon at the Syzran deposit did not use butyl cyclohexane, hexadecane, or paraffin (Shturm, 1951).

Another sspect of this work involves an explanation of the effect of facultative-anaerobic heterotrophic bacteria, without the admission of air, on oil. The decomposition of oil under these conditions (Ékzertsev, 1956) by microflora from productive horizons was accompanied by the elimination of gas consisting of methane (14-35% by volume), carbon dioxide (1.9-5.0%), hydrogen (4.4-6.2%), and nitrogen (61-78.1%). Similar results were obtained by decomposition of Sakhalin oil (Bokova, 1950).

A number of other papers discuss the decomposition of oil in a nitrogen atmosphere (Simakova, 1956; Simakova and others, 1958; Bolotskaya and others, 1957; etc.).

Conclusions

1. Temperatures that do not exceed the optimum values characteristic of the various species of microorganisms favor the growth of these organisms. Because of the high thermal tolerance of thermophyles, which are found

in deep sedimentary strata, microbiological processes may take place in deep horizons and formational waters at relatively high temperatures.

2. An increase in external pressure (such as hydrostatic) retards denaturation and favors enzymatic activity at temperatures above the optimum.

3. Bacteria vary greatly in their ability to transmit extreme osmotic pressures. This capacity depends on the salinity of the medium in which the organisms live in nature.

4. The effect of free hydrogen ions on microorganisms depends not only on the degree of dissociation of the H-bearing electrolyte, but also on the presence of salts, which control the buffering and oxidation-reduction properties of the medium. The relation of microorganisms to pH and Eh also depends on the natural conditions under which the organisms live.

5. The actual values of some basic ecological factors—temperature, pressure, salinity, pH—are such that microbiological processes are possible in oil deposits.

LITERATURE CITED

Allen, M. B. 1950. "The dynamic nature of thermophily." J. Gen. Physiol., 33, No. 3, p. 205.

Baars, J. K. 1930. "Over sulfaatreductie door bacterien." Dissertation, Delft.

Baas-Becking, L. G. M. and Kaplan, J. R. 1956. "Biological processes in the estuarine environment. III. Electrochemical considerations regarding the sulfur cycle." Proc. Konikl. Nederl. Akad. wet. C., p. 86-96.

Basset, J. and Macheboeuf, A. 1931. "Etude sur les effects biologiques des ultrapressions. Resistance des bactéries, des diastases et des toxines, et pressions tres élevées." C. r., 195.

Bastin, E. S. 1926. "The presence of sulfate-reducing bacteria in oil field waters." Science, 63, No. 1618, p. 21.

Beerstecher, E. 1954. Petroleum Microbiology, Van Nostrand, Princeton.

Bokova, E. N. 1950. "The formation of methane during microbiological decomposition of petroleum." Field and Industrial Geochemistry [in Russian], No. 2, Gostoptekhizdat.

Bolotskaya, O. P., Shmonova, N. I., and Strigalova, N. V. 1957. Nature of Changes in Oil under Anaerobic Conditions and the Biogenic Factor [in Russian], Collection of Author's Abstracts, VNIGRI.

Butlin, K. R. and Postgate, J. R. 1953. "Microbiological formation of sulfide and sulfur. Symposium metabolismo microbica, Roma, p. 127-142.

Campbell, L. L, Frank, H. A., and Hall, E. R. 1957. "Studies on thermophilic sulfate-reducing bacteria, I. Identification of Sporovibrio desulfuricans as Clostridium nigrificans." J. Bact. 516-521.

Crozier, N. O., Johnson, F. H., and Brown, D. 1942. Science, 95, No. 2460, p. 200.

Davis, J. B. 1956. "Microbial decomposition of hydrocarbons." Industr. and Engin. Chem. 48, No. 9, p. 1444-1448.

Delden, V. van. 1904. "Beitrag zur Kenntnis der Sulfatreduktion durch Bakterien." Cbl. Bakt. Abt. II, Bd. 2, S. 117.

Ékzertsev, V. A. 1956. Dissertation: Formation of Methane by Microorganisms in Oil Deposits [in Russian], OBN AN SSSR.

Eyring, H. and Maggee, I. Z. 1942. "Application of the theory of absolute reaction rates to bacterial luminescence." Cellular and Compar. Physiol., 20, p. 169-177.

Ginzburg-Karagivheva, T. L. 1926. "Microbiological investigation of the hydrogen-sulfide waters of Apsheron," Azerb. neft. khoz-vo, No. 6-7.

Isachenko, B. L. 1927. "Microbiological investigations on mud lakes," Trudy Geol. kom., novaya seriya, No. 148, Leningrad.

Johnson, F. H., Brown, D., and Marsland, D. A. 1942. "A basic mechanism in the biological effects on temperature, pressure, and narcotics." Science, 95, No. 2460, p. 200-203.

Johnson, F. H. and Lewin, J. 1946. "The influence of pressure, temperature, and quinine on the rates of growth and disinfection of E. coli in the logarithmic growth phase." J. Cellular and Compar. Physiol. 28, p. 77-97.

Kaplan, J. R. 1956. "Evidence of microbiological activity in some of the geothermal regions of New Zealand." N. Z. J. Sci. and Technol., 37, p. 639-662.

Khait, S. Z. 1924. "Microspira desulfuricans in the Kuyal'nitskii Estuary." Zhurn. n.-i. kafedr v Odesse, 1, No. 10-u.

Kolesnik, Z. A. 1955. Microflora of Oil Deposits in the Terek-Dagestan Oil District [in Russian], Geological Collection of VNIGRI, No. 1, Gostoptekhizdat.

Kuznetsova, V. A., Ashirov, K. B., Gromovich, V. A., Ovchinnikova, I. V., and Kuznetzov, S. K. 1957. "An experiment to suppress the growth of sulfate-reducing bacteria in the oil bed of the Kalinovka field." Mikrobiologiya, 26, No. 3.

Maliyants, A. A. 1933. "Microbiological investigations of Caspian Sea muds." Trudy Azerb. neft. issl. inst., Azneftizdat.

Militzer, W. E., Sonderegger, T. B., Tuttle, L. C., and Georgi, C. E. 1950. "Thermal enzymes. II. Cytochromes." Arch. Biochem. 26, pp. 299-306.

Miller, L. P. 1949. "Rapid formation of high concentrations of hydrogensulfide by sulfate-reducing bacteria." Contrib. Boyce Thompson Inst., 15, p. 437.

Miller, L. P. 1950. "Formation of metal sulfides through the activities of sulfate-reducing bacteria." Contrib. Boyce Thompson Inst. 1950, 16, p. 85.

Novelli, G. D. and ZoBell, C. E. 1944. "Assimilation of petroleum hydrocarbons by sulfate-reducing bacteria." J. Bact. 47, pp. 447-448.

Pel'sh, A. D. 1937. "Dynamics of the desulfatization process." Trudy Vses. inst. galurgii, No. 14, Izd. AN SSSR.

Postgate, J. R. 1959. "Some problems in the field of bacterial sulfate-reduction." Ann. Rev. Microbiol. 13 (in press).

Prévot, A. R., Raynaud, M., and Tatoki, M. 1951. "Recherches sur la thermorésistance de Cl. sporogenes et le phénomène d'entrainement des espèces peu résistantes." Ann. Inst. Pasteur., 80, p. 533.

Rabotnova, I. L. 1957. The Role of Physicochemical Conditions (pH and rH$_2$) in the Life Activity of Microorganisms [in Russian], Izd. AN SSSR, Moscow.

Rittenberg, S. C. 1941. "Studies on marine sulfate-reducing bacteria." Cited by ZoBell, 1958.

Rosenfeld, W. D. 1947. "Anaerobic oxidation of hydrocarbons by sulfate-reducing bacteria." J. Bact. 54, pp. 664-665.

Rubenchik, L. I. and Goikherman, D. G. 1936. "The effect of freshening of estuaries on the microflora of medicinal muds." Arkhiv biol. nauk, 43, No. 2-3.

Rubenchik, L. I. and Goikherman, D. G. 1940. "The microbiology of mud lakes, III: Vertical distribution of bacteria in the bottom deposits of Repnoe Lake." Mikrobiologiya, 9, No. 1.

Shturm, L. D. 1951. "The role of sulfate-reducing bacteria in the life and history of oil deposits." from the Collection devoted to the memory of M. I. Gubkin [in Russian], Izd. AN SSSR, Moscow.

Simakova, T. L. 1956. "Bacterial factor in the alteration of oil and the transformation of initial organic material." Author's abstract of scientific papers, VNIGRI, No. 17.

Simakova, T. L., Gorskaya, A. I., Kolesnik, Z. A., Bolotskaya, A. P. Strigalova, N. V., and Shmonova, I. P. 1958. "Nature of the alteration of oil by the biogenic factor." Trudy VNIGRI, No. 128.

Starkey, R. L. and Wight, K. M. 1945. Anaerobic Corrosion of Iron in Soil. Techn. Rep. Distrib. Amer. Gas. Assoc. Communication, No. 945, p. 108.

Updegraff, D. M. and Wren, G. B. 1954. "The release of oil from petroleum-bearing materials by sulfate-reducing bacteria." Appl. Microbiol., 2, No. 6, p. 309.

Uspenskii, V. A., Gorskaya, A. I., and Karpova, I. P. 1947. "The origin of algarites and the processes of anaerobic oxidation of oil." Izv. AN SSSR, seriya, geol., No. 4.

ZoBell, C. E. 1946a. Marine microbiology, Baltimore.

ZoBell, C. E. 1946b. "Studies on redox potential of marine sediments." Bull. Amer. Assoc. Petrol. Geologists., 30, pp. 477-513.

ZoBell, C. E. 1947. "Bacterial release of oil from sedimentary materials." Oil and Gas. J., 46, pp. 62-65.

ZoBell, C. E. 1950. "Assimilation of hydrocarbons by microorganisms." Advances Enzymol., 10, p; 443.

ZoBell, C. E. 1958. "Ecology of sulfate-reducing bacteria." Producers Monthly, 22, pp. 12-20.

ZoBell, C. E. and Johnson, F. W. 1949. "The influence of hydrostatic pressure on the growth and viability of terrestrial and marine bacteria." J. Bact. 57, pp. 172-189.

ZoBell, C. E. and Oppenheimer, C. H. 1950. "Some effects of hydrostatic pressure on the growth, multiplication and morphology of marine bacteria." J. Bact. 60, pp. 771-781.

THE ROLE OF MICROORGANISMS IN THE FORMATION
AND DESTRUCTION OF SULFUR DEPOSITS

M. V. Ivanov

(Institute of Microbiology, Academy of Sciences, USSR, Moscow)

The question of participation of microorganisms in the formation of sulfur deposits was raised in the geological literature nearly 50 years ago. Basing their theories on the works of Beijerinck (1895) on the physiology of Spirillum desulfuricans, the geologists Stutzer (1911) and Hunt (1915) proposed that these microorganisms took part in the formation of the Sicilian sulfur deposits. Vernadskii understood the significance of microorganisms in the processes of sulfur formation more clearly than other geologists. Being thoroughly acquainted with the works of the microbiologists Vinogradskii, Omelyanskii, Isachenko, and Waksman, Vernadskii stated directly (1927): "The biochemical precipitation of native sulfur obviously defines the special character of the distribution of sulfur deposits in the biosphere. These deposits are related to the physical-geographic and ecological conditions, and are the consequence of the distribution of sulfur-precipitating organisms" (p. 333). Under the influence of Vernadskii's ideas, geologists occupied in our country in the thirties with the problem of discovering the origin of sulfur did much to establish the biogenic theory of sulfur formation (Danov, 1936; Drobyshev, 1930; Murzaev, 1939).

An almost complete absence of actual data on the distribution of microorganisms in sulfur-bearing deposits until recently compelled geologists to turn to information of microbiologists, obtained from studies of basins and formational waters of oil fields.

The incomplete development of the microbiological side of the theory of sulfur origin led some geologists (Uklonskii, 1940; Andreev, 1955) to deny completely the participation of microorganisms in the formation of sulfur and to attempt to treat the formation and oxidation of hydrogen sulfide and sulfur from a purely chemical point of view (so-called mineral reduction).

Only in the past 10-12 years have there appeared in domestic and foreign geochemical and microbiological literature rather numerous data demonstrating the role of microorganisms in the formation of deposits of native sulfur.

In a methodical approach to the problem, these papers may be divided into two groups. One group of papers represents the work of microbiologists, the purpose of which is the study of the distribution of microorganisms directly in sulfur-bearing sedimentary rocks. Attention is devoted chiefly, among detailed investigations, to sulfate-reducing and sulfur-oxidizing bacteria, since sulfur deposits are genetically related to deposits of sulfate minerals, chiefly gypsum (Vernadskii, 1927; Danov, 1936; Sokolov, 1958a and 1959b), and native sulfur forms by oxidation of hydrogen sulfide, which in turn is a product of reduction of sulfates. Results of studying the distribution of sulfate-reducing and sulfur-oxidizing bacteria such as Thiobacillus thioparus in rocks of sulfur deposits in the Soviet Union, Italy, India, Austria, and the United States of America are shown in Table 1, from which it may be seen that bacteria of the sulfur cycle are very widespread in sedimentary deposits of sulfur.

However, among proofs of the participation of microorganisms in geological processes, the mere record of the presence of any particular physiological group is but one, and by no means the most convincing, argument. Actually, in sedimentary rocks and in formational waters, microorganisms are generally found under special conditions, having nothing or almost nothing in common with conditions that we create in nutrient media used for counting and for culturing bacteria in the laboratory.

TABLE 1. Distribution of Sulfate-Reducing and Sulfur-Oxidizing Bacteria in Sedimentary Rocks in the Waters of Various Native-Sulfur Deposits

Deposit	Brief description of investigated samples	Bacteria		Author, year
		sulfate reduc- ing	sulfur oxid- ixing	
Chekur-Koyash (Crimea)	Seams of sulfur and gypsum in clayey marl	+	+	Shturm and Simakova, 1928
Krasnovodsk (Turkmenia)	Dispersed sulfur in Akchagylian deposits	+	+	Shturm and Simakova, 1928
Coast of Bay of Bengal (India)	Sulfur in Recent clay sediments	+	−	Subba Rao, 1949
Salt domes (U. S. A.)	Caprock of salt domes	+	−	ZoBell, 1947
Sicily	Sulfur-bearing marls and clays	+	+	Miller, 1949
Alekseevka (Kuibyshev Oblast)	Sulfur in bituminous limestone	+	+	Schwartz, 1958
Vodino (Kuibyshev Oblast)	Sulfur-bearing clays and limestones	−	+	Shturm, 1957 Ivanov, 1957c
Lake Eyre (Australia)	Nodules of sulfur in lower Quaternary deposits	+	+	Ivanov, 1957c Ivanov, 1957c
Shor-Su (Uzbek SSR)	Sulfur-bearing limestones and marls, ground water, deposits of sulfur	+	+	Baas-Becking and Kaplan, 1957
Lyuben'-Velikii	Hydrogen sulfide, calcium sulfate ground water, sulfur-bearing limestone and underlying gypsum-anhydrite.	+	−	Ivanov (data collected in 1958-59)
Nemirov Lvov Oblast		+	+	
Rozdol Ukraine SSR		+	+	
Yazov		+	+	

Explanation: + bacteria detected, − no analysis made.

A high and frequently very distinctive mineralization of ground water, the presence of quite special forms of organic material, the presence of distinctive biocoenoses (historically made up of microbes) with the entire complex range of antagonistic and symbiotic relations of organisms, and the commonly elevated radioactivity and temperature — these represent far from a complete list of the specific ecological conditions under which microorganisms are found in sedimentary rocks and about which we still know very little.

In view of all this, it is necessary to realize that the skeptical attitude of some geologists to proofs of geochemical activity of microorganisms under conditions found in sedimentary rocks has some real basis. Microbiologists must study the actual ecological conditions under which the microorganisms exist and must attempt to determine the rate of microbiological processes under natural conditions.

The necessity of investigating the rate of microbiological processes under natural conditions has been mentioned repeatedly in microbiological literature (Vinogradskii, 1924, 1945; Isachenko, 1951; Khudyakov, 1954), but until recently proper methods of treating this problem had not been developed.

A certain progress in this direction appeared only after the extensive introduction of the highly sensitive method of radioactive isotopes as an investigatory procedure.

The value of this method lies in the fact that it permits one to define the intensity of a natural process under conditions of a short-lived experiment directly on natural material (mud, lake water, formational water). We may illustrate this method by the limits of the intensity of sulfate reduction (Ivanov, 1956). A solution of $Na_2S^{35}O_4$ is added slowly to test samples of formational waters after they have been taken from the hole or from a spring. The flask with the water stoppered with a ground-glass stopper and allowed to stand for a period no longer than one or two days, preferably at a temperature corresponding to the temperature of the water being studied. During the entire course of this short-period experiment the sulfate-reducing bacteria in the formational water remain in their natural hydrochemical and biological environment; they utilize the same organic mineral substances and, what is especially important, in the same concentrations that obtain in the investigated water.

At the end of the experiment, the radioactivity of the remaining sulfates and the newly formed hydrogen sulfide in the water is determined, as well as the total quantity of sulfates, by the formula

$$x = \frac{AH_2S \cdot S/SO_4}{ASO_4},$$

where AH_2S represents the radioactivity of H_2S in impulses per minute per liter, ASO_4 the radioactivity of SO_4 in impulses per minute per liter, S/SO_4 the content of sulfur in sulfates in milligrams per liter, and x the quantity of H_2S that formed during the course of the experiment, in milligrams per liter.

The results of using tracer compounds of sulfur for clarifying some of the details of the origin of different sulfur deposits will be discussed somewhat later. Here we wish merely to mention that by using such methods we have succeeded in proving beyond doubt that sulfate-reducing and sulfur-oxidizing bacteria fully preserve their life activity under conditions obtaining in the ground waters of oil and sulfur deposits, and they vigorously participate in transforming sulfur compounds (Ivanov, 1957a, 1957b, 1957c, 1958, 1959).

TABLE 2. Ratio of Stable Sulfur Isotopes S^{32}/S^{34} in Natural Sulfur Compounds of Biogenic and Abiogenic Origin

Compounds not aided by microorganisms in their formation	S^{32}/S^{34}
Sulfates of gypseous rocks	20.83-22.10
Sulfates of ground water	21.16-22.20
Sulfates of sea water	21.70-21.84
Sulfur from volcanic deposits	21.82-22.10
Compounds formed through the participation of sulfate-reducing bacteria	
Sulfur from sedimentary deposits	21.80-22.70
Hydrogen from ground water	21.50-22.70

Thus, the investigations on microflora in sulfur-bearing rocks and on the intensity of microbiological processes directly in sulfur deposits, made by microbiologists in recent years, undoubtedly point to the correctness of the basic position of the biogenic theory concerning the origin of sulfur deposits (a theory formulated as early as the beginning of the twentieth century in the papers of Vernadskii and Hunt).

The papers of geochemists on the distribution of stable isotopes of sulfur in natural compounds also point to the exceptional importance of microbiological processes in the origin of sulfur deposits. In mass-spectrographic analysis of the content of the stable isotopes S^{32} and S^{34} in various compounds, geochemists have discovered that the ratio S^{32}/S^{34} in the hydrogen sulfide of formational water and in the sulfur of sedimentary deposits differ rather markedly from the ratio of isotopes in volcanic sulfur and sulfates (Table 2) (Ingerson, 1954; Thode, Macnamara, and Fleming 1954; Rankama, 1956). To explain this phenomenon it has been suggested that the separation of stable isotopes of sulfur occurs during microbiological reduction of sulfates to hydrogen sulfide, and that just compounds of the first group (sulfur and sulfides of sedimentary origin, and hydrogen sulfide) have been relatively enriched in S^{32} and impoverished in S^{34}.

An experimental varification of this proposal, undertaken by Thode and his co-workers in 1951, completely confirms the capacity of sulfate-reducing to reduce the light isotope of sulfur S^{32} more energetically (Thode, Kleerkoper, and McElcheran, 1951). Further experiments with cultures of sulfur-reducing bacteria, carried on by various groups of American microbiologists, including a group under the direction of Starkey, have also demonstrated that hydrogen sulfide formed by sulfate-reducing bacteria differs from the initial sulfates by a higher S^{32}/S^{34} ratio (Jones and Starkey, 1957; Jones, Starkey, Feely, and Kulp, 1956). Data from a typical experiment on separating sulfur isotopes while growing cultures of sulfate-reducing bacteria are shown in Fig. 1 (Feely and Kulp, 1957).

According to Rankama (1956), the accumulated data on distribution of sulfur isotopes in natural compounds permit one to assert that, if the ratio S^{32}/S^{34} in an investigated compound of sulfur is greater than 22.3, this compound is of biogenic origin. As may be seen from the data in Table 3, most of the sedimentary deposits of sulfur thus far investigated have a "biogenic" ratio of sulfur isotopes.

TABLE 3. Ratio of Stable Sulfur Isotopes S^{32}/S^{34} in Native Sulfur and Sulfates from Various Sedimentary Deposits of Sulfur

Deposit	Sulfur	Sulfate	Author, year
Sicily	22.03	—	Ingerson, 1954
Cyrenaica	22.57	21.88	Macnamara and others, 1951
Coast of Gulf of Mexico	22.29 22.37	—	Thode and others, 1951
Salt Domes, U. S. A.	21.86 22.44	20.83 21.93	Feely and Kulp, 1957

Thus, all the enumerated data, obtained by collaboration of microbiologists and geochemists, permit one to state that sulfur in sedimentary deposits is a product of the activity of microorganisms.

An investigation of the distribution and activity of microorganisms participating in the sulfur cycle in present-day lakes and formational water allow us also to discern the roles of the individual physiological groups of bacteria in the formation of sulfur deposits of various types.

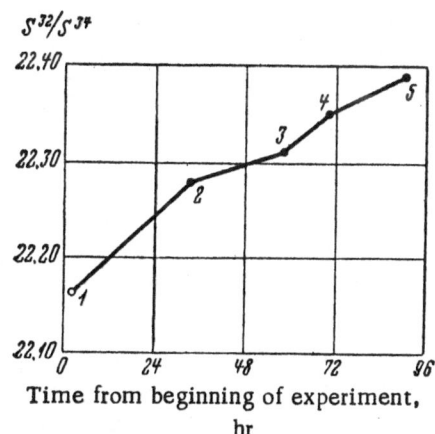

Time from beginning of experiment, hr

Fig. 1. Ratio of sulfur isotopes S^{32}/S^{34} in initial sulfate (1) and in hydrogen sulfide, formed by reduction of this sulfate in a culture of sulfate-reducing bacteria at various times after the beginning of the experiment (2, 3, 4, 5). From the data of Feely and Kulp (1957).

In the geologic literature all sedimentary deposits of sulfur are divided into two groups: syngenetic deposits, forming in the bottom sediments of hydrogen-sulfide basins simultaneously with the enclosing sediments, and epigenetic deposits, in which the sulfur is deposited in previously formed sedimentary rocks by secondary oxidation of hydrogen-sulfide ground water.

The Black Sea, the lower layers of which are contaminated with hydrogen sulfide, is generally cited as representing the syngenetic conditions of sulfur accumulation described in the geologic literature (Danov, 1936, Murzaev, 1939). However, this example is clearly unfortunate, since the content of H_2S in the bottom water of the Black Sea is small, and all the processes of the sulfur cycle in the water of this sea probably occur at a very slow rate. It would be much better to examine these processes in hydrogen-sulfide lakes, where the concentration of H_2S in the water commonly exceeds 100 mg/liter and in the bottom deposits reaches 1000 mg/liter and more (Kuznetsov, 1952). The origin of this hydrogen sulfide has been rather well studied by Isachenko (1927), Rubenchik (1949), and Kuznetsov (1952); these authors have shown that most of the hydrogen sulfide forms by microbiological reduction of sulfates, chiefly in the bottom sediments.

However, until recently there were no quantitative data concerning the formation of hydrogen sulfide in aqueous basins.

The use of tracer elements has demonstrated that in some muds the rate of H_2S formation is measured in tens of milligrams per liter per day, and is thus fully equivalent to the rate of the process under laboratory conditions (Table 4). By this method it has been shown that the most intense generation of hydrogen sulfide occurs in the uppermost layer of the bottom sediments, the rate decreasing rapidly with depth in these sediments, although the quantity of sulfate-reducing bacteria in the deeper muds is by no means less than in the surface layer (Ivanov and Terebkova, 1959a; Sokolova and Sorokin, 1958).

TABLE 4. Intensity of Hydrogen Sulfide Formation in mg/liter in Surface Layer of Bottom Mud per Day

Basin	H_2S	Author, year
Lake Belovod'	0.067-0.127	Ivanov, 1956
Lake Solenoe (winter)	0.388-1.404	Ivanov and Terebkova, 1959a
Lake Solenoe (summer)	0.860-18.97	Ivanov and Terebkova, 1959b
Lake Beloe	0.0002-0.0011	Sokolova and Sorokin, 1957
Rybinskoe Reservoir	0.0006-0.092	Sokolova and Sorokin, 1957
Gor'kovskoe	0.041-2.943	Sokolova and Sorokin, 1958
Littoral zone of Barents Sea	0.68-24.26	Ivanov and Ryzhkova, 1960

The activity of bacteria that oxidize hydrogen-sulfide to sulfur has been studied less thoroughly. It is clear that the group of achromatic sulfur bacteria cannot play any substantial role in the accumulation of sulfur, since representatives of this group grow only in water with low concentrations of hydrogen sulfide.

In the microbiological literature of recent years there have been numerous references to the participation of colored sulfur bacteria in the oxidation of hydrogen sulfide to sulfur in lakes. Thus, Butlin and Postgate have discovered intense growth of the bacteria Chromatium and Chlorobium in three sulfur-producing lakes in Cyrenaica. The mud deposits of these lakes consist almost entirely of molecular sulfur. On the basis of these observations, and also of laboratory experiments (in which molecular sulfur was obtained from a mixed culture of sulfur-reducing bacteria and colored sulfur bacteria), it has been concluded that these two groups of bacteria are extremely important in the formation of sulfur (Butlin and Postgate, 1952, 1953). A role no less important in the oxidation of hydrogen sulfide is played by purple sulfur bacteria in biologically anisotropic lakes, where these organisms occur in quantities as great as five million cells per milliliter at the boundary of the hydrogen-sulfide zone (Kuznetsov, 1952). The intensity of bacterial photosynthesis of the purple sulfur bacteria in Belovod' Lake reaches 1.2 mg of CO_2 per liter of water per day (Lyakova, 1957), and the amount of molecular sulfur deposited by these organisms is as much as 1.7 mg S per liter.[*]

A direct proof of activity of sulfate-reducing bacteria directly in formational water is found in the results of experiments with tracer sulfate, as shown in Table 5. From the data in this table it may be seen that very active microbiological reduction of sulfates occurred in all the investigated samples; of these the ground waters of Rozdol and Yazov represent weakly mineralized near-surface waters, whereas the Shor-Su sample represents sulfate-reduction in typical sodium-chloride brines with mineralizations of 80-100 g/liter.

Recently the question of the source of the organic material used by sulfate-reducing bacteria in ground water has been clarified to a certain extent. Although there are no reliable data in the literature on the use of oil by pure

---

[*] We have computed the amount of deposited sulfur from Lyalikova's data (Lyalikova, 1957) by means of the balanced equation of bacterial photosynthesis: $CO_2 + 2H_2S \rightarrow (CH_2O) + 2S$.

cultures of these bacteria, there is no doubt that in a mixed culture hydrogen sulfide forms sulfates in a mineral medium containing oil as the only source of organic material (Kuznetsova and others, 1957; Gasanov, 1961; Malyshek and Gasanov, 1959; Simakova, 1961).

Microflora causing oxidation of hydrogen sulfide in ground water have been investigated to a lesser degree. Achromatic and colored sulfur bacteria are apparently not widespread in ground water. A single reference in the literature to massive development of purple sulfur bacteria in the formational water of the Surukhan oil field (Malyshek and Maliyants, 1935; Malyshek, Maliyants, and Reinfel'd, 1935) is the sole cited example, and this remains an unsolved puzzle. Sulfur-oxidizing bacteria such as Thiobacillus thioparus have been rather frequently observed, first during the investigations of Ginzburg-Karagicheva (1926) in hydrogen-sulfide ground water and in a deep core. Shturm (Shturm and Simakova, 1928; Shturm, 1937) and, later, other investigators (see Table 1), found these bacteria in sulfur-bearing rocks, but until recently the problem of their participation in the sulfur cycle in ground water has remained uncertain. The fact is that all sulfur-oxidizing bacteria except Thiobacillus denitrificans, have been considered strictly anaerobic forms, and their presence of hydrogen-sulfide waters and muds has not been explained.

The investigations of Baalsrud and Baalsrud and of De Kruyff and his co-workers indicate, however, that different species of Thiobacteria are rather easily converted from aerobic to anaerobic forms by using nitrates according to the process of sulfur denitrification (Baalsrud and Baalsrud, 1954; De Kruyff, Van der Walt, and Schwartz, 1957).

Later data of Sokolova show that the most intensive growth of Thiobacillus thioparus in a medium with calcium sulfide occurs at an rH_2 of 12-16 (Kuznetsov and Sokolova, 1960).

TABLE 5. Intensity of Microbiological Reduction of Sulfates in Ground Waters of Sulfur Deposits in the USSR

Deposit	Dry residue, g/liter	H_2S, mg/liter	Intensity of H_2S formation, mg/liter per day
Shor-Su	86.6	1627	0.078
Shor-Su	102.8	482	0.179
Nemirov	1.9	130	0.204-0.260
Lyuben'	1.7-1.9	70-94	0.547
Rozdol	1.6-2.3	23.65	0.129-2.015

In connection with Sokolova's laboratory experiments, it is of interest to note that, under natural conditions, deposits of sulfur form at the same oxidation-reduction potential (Table 6) by oxidation of hydrogen-sulfide water.

A schematic view of the participation of microorganisms in the formation of sulfur deposits at Shor-Su is shown in Fig. 2. Under the conditions of low oxidation-reduction potential existing at the oil deposit on anticline IV and in the lower horizons of anticline II, sulfates are reduced to hydrogen sulfide; this process is especially active in the Bukhara and Alai strata, which contain significant stores of gypsum. At anticline IV, where lower Paleogene rocks are covered by a thick sequence of clays, hydrogen sulfide is conserved to a considerable degree, and only part of it is diffused to the earth's surface, where small sulfur deposits have formed.

At the second anticline, where lower Paleogene rocks are exposed, deep waters are diluted to some extent by surface waters, the mineralization drops, the potential increased to 12-14 rH_2 units, and oxidation of hydrogen sulfide to sulfur begins in the water through participation of sulfur-oxidizing bacteria (Ivanov, 1958).

With gradually increasing pressure in the gas at anticline IV, the level of hydrogen-sulfide water at anticline II gradually rose, and the sulfur already being deposited found itself in a reducing environment that excluded oxidation of the sulfur.

TABLE 6. Distribution of Thiobacteria Depositing Sulfur by Oxidation of Hydrogen Sulfide in Waters at Various Sulfur Deposits of Shor-Su

Sample locality	pH	rH$_2$	H$_2$S, mg/liter	No. of thio-bacteria per ml	Deposit of sulfur
Spring 40, bed K	7.45	19.25	0	0	−
Hole 51, bed K	7.40	10.66	29.0	10	+
Horizon III, bed K	7.00	12.10	62.7	1000	+
Horizon III, bed M	7.35	15.70	236.0	100	+
Horizon V, bed K	7.30	14.30	35.0	1000	+
Spring, bed M	6.80	7.90	275.0	10	−
Hole, bed M	7.00	7.60	987.0	0	−
Spillway 6	7.60	14.40	12.0	100	+
Spillway 7	7.40	17.40	0	10	−
Creek with H$_2$S	7.60	12.44	63.1	100	+
Creek without H$_2$S	7.20	19.05	0	10	−

Explanation: + sulfur deposit, − no sulfur.

The participation of microorganisms in the formation of epigenetic salt-dome deposits in Texas and Louisiana has been demonstrated in the paper of Feely and Kulp (1957) by the distribution of stable sulfur isotopes in various rocks forming deposits of this type. The data of these investigators are shown in Table 7. We see that in the non-sulfurous rocks of the salt dome the ratio S^{32}/S^{34} ranges between the very narrow limits (21.82-21.90), similar to the ratio in normal sedimentary rocks and in sea water. The insignificant variation in the data obtained attests that reduction of sedimentary rocks did not take place in the salt dome. In the slightly sulfurous anhydrite caprock, the sulfates were to a considerable degree subjected to microbiological reduction, as indicated by the ratio of sulfur isotopes, equal to 21.64 in several samples. Reduction was even more active at the base of the sulfur-bearing bed, in the calcitic caprock, where the ratio S^{32}/S^{34} for some samples of sulfates is 20.8.

The impoverishment of the sulfates in the light isotope of sulfur by reduction through microorganisms led to an enrichment of the molecular sulfur in the isotope S^{32}, as is clearly seen also in Table 7.

The activity of sulfate-reducing bacteria apparently does not die out even after the sulfur-bearing deposits are covered by younger sedimentary rocks. It is no accident that almost all deposits of sulfur accompany hydrogen-sulfide water, which is an indirect characteristic utilized in searching for sulfur (Sokolov, 1958a and 1958b).

Close agreement between boundaries of sulfur distribution and boundaries of underground basins of hydrogen-sulfide water is especially well marked in the deposits of the Carpathian region, and this compels us to believe that the phenomena are mutually related. Our observations on the Nemirov, Lyuben', Yazov, and Rozdol deposits have shown that sulfate-reducing bacteria are widely distributed in the ground water about sulfur deposits; these organisms, as experiments with tracer sulfate have shown, actively form hydrogen sulfide and, even more, produce a low oxidation-reduction potential (to an rH$_2$ of 7-9), which hinders the rise of oxidation processes (Ivanov, 1959).

In the light of these data the restriction of the Carpathian deposits to the zone joining the platform structure to the Carpathian marginal depression may be considered not only from the viewpoint of the origin of these deposits through reduction of sulfates by bases migrating from the depression (Sokolov, 1958a), but also from the viewpoint of development of zones of restricted water exchange, in the districts adjoining the depression, where conditions are favorable for the growth of sulfate-reducing bacteria, the activity of which fosters the preservation of previously deposited accumulations of sulfur.

When sulfur deposits are found above the level of hydrogen-sulfide waters, either through natural tectonic processes and erosion or through the economic enterprises of man, oxidation of the sulfur begins and alum or other secondary sulfate "caps" are formed (Uklonskii, 1940). Oxidation is especially strong at deposits where sulfur-bearing beds are exposed by natural processes (Gaurdak, Shor-Su, Kara-Kumy). At such places it is common for the sul-

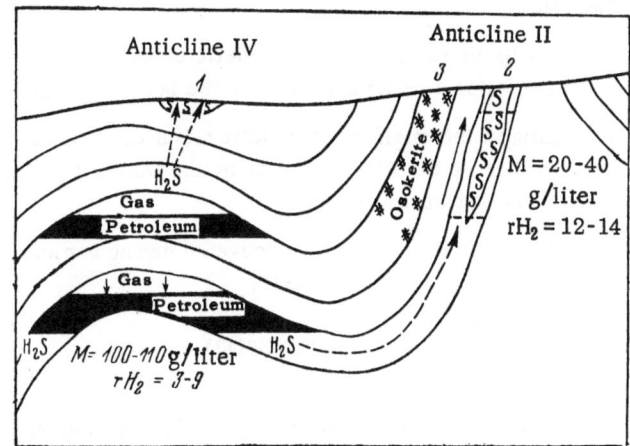

Fig. 2. Diagram showing formation of sulfur deposit at Shor-Su. M) Amount of dry residue in water in g/liter; rH_2) value of oxidation-reduction potential; 1) small sulfur deposit in dome of anticline IV; 2) principal sulfur deposit in eroded structure of anticline II; 3) deposit of ozocerite.

fur to be completely oxidized; the surrounding rock is strongly altered by sulfuric acid that has formed, and the pH of samples, even in carbonate rocks, drops below 1.0 (Ivanov, Lyalikova, and Kuznetsov, 1958).

Oxidation of sulfur begins rather quickly also in artificially exposed quarries of sulfur-bearing rocks. Thus, at the Vodino deposit, in a 15-year-old quarry, the sulfur-bearing rocks had a distinctly acid reaction, and the sulfur easily crumbled in the hand (Ivanov, 1957c).

TABLE 7. Ratio of Stable Isotopes of Sulfur S^{32}/S^{34} in the Section of a Sulfur-bearing Salt Dome

Rock	Sulfate	Molecular sulfur
Calcitic caprock, principal deposit of sulfur	20.80-21.93	21.84-22.44
Slightly sulfurous anhydritic caprock	21.64-21.95	22.06-22.17
Rocks of nonsulfurous salt mass	21.82-21.90	None

The data of Karavaiko (1960) on the Rozdol deposit show that at a certain pH in the rocks, features of incipient sulfur oxidation may be detected even earlier, not in the gross sample, but in microscopic preparations and in impressions on indicator paper, as early as two or three years after lowering of the hydrogen-sulfide water level.

The investigation of microflora that cause oxidation processes in sulfur-bearing rock, an investigation carried out in recent years at the Institute of Microbiology, has shown that the sulfur-oxidizing bacteria Thiobacillus thiooxidans is present in large quantities in all samples of oxidized sulfur ores; this organism, as is known, is capable of active oxidation of sulfur and it withstands very high concentrations of sulfuric acid.

The investigation of the geochemical activity of Thiobacillus thiooxidans is continuing at the present time. The question arises of developing a method for determining the intensity of this process under natural conditions;

however, the data obtained permit us to state that the geologically short-lived existence of sulfur deposits, as noted by Vernadskii (1912-1922), is explained in great measure by the activity of sulfur-oxidizing thiobacteria.

Conclusions

1. The principal organisms taking part in the formation of sedimentary sulfur deposits are the sulfate-reducing and the sulfur-oxidizing bacteria.

2. The role of microorganisms in the formation of syngenetic deposits at the present time is demonstrated in several hydrogen-sulfide lakes, where sulfur is being deposited in the bottom sediments.

3. The participation of bacteria in the origin of epigenetic sulfur deposits has been demonstrated by direct microbiological experiments (at the Shor-Su deposit) as well as by the works of geochemists on the distribution of the stable sulfur isotopes (American deposits at salt domes).

4. At sulfur deposits exposed by natural processes or uncovered during exploitation, sulfur begins to oxidize to sulfuric acid through participation of sulfur-oxidizing bacteria.

LITERATURE CITED

Andreev, P. F. 1955. "The bacterial hypothesis of the origin of oil types." Geological Collection of VNIGRI, No. 1.

Andreevskii, I. L. 1961. "The effect of microflora in the third bed of the Yarega deposit on changes in composition and properties of the oil." Present volume, see p. 56.

Baalsrud, K. and Baalsrud, K. S. 1954. "Studies of Thiobacillus denitrificans." Arch. Microbiol. 20, H. 1, p. 34.

Baas-Beking, L. G. M. and Kaplan, I. R. 1957. "The microbiological origin of the sulfur nodules of Lake Eyre." Trans. Roy. Soc. S. Austral., 79, p. 52.

Balasundaram, M. S. 1954. "Occurrence of sulfur near Kona, Krishna District, Madras." Indian Minerals, 8, No. 2, p. 102.

Beerstecher, E. 1954. Petroleum Microbiology, Van Nostrand, Princeton.

Beijerink, M. W. 1895. "Über Spirillum desulfuricans als Ursache von Sulfatreduktion." Cbl. Bacteriol. Abt. VI, Bd. 1.

Butlin, K. R. and Postgate, J. R. 1952. "The microbiological formation of sulfur in Cyrenaican lakes." Biology of Deserts. Inst. of Biol., London, p. 112.

Butlin, K. R. and Postgate, J. R. 1953. "Microbiological formation of sulfide and sulfur." Sympos. Microbiol. Metabolism, Roma, 126.

Danov, A. V. 1936. Conditions of Forming Sulfur Deposits in Central Asia [in Russian], Izd. ONTI NKTP.

De Kruyff, C., Van der Walt, J., and Schwartz, H. 1957. "The utilization of thiocyanate and nitrate by Thiobacilli." Ant. van Leeuwenhoek, 23, p. 305.

Drobyshev, D. 1930. "The origin of sulfur deposits in mountainous Dagestan." Trudy Geolkoma, No. 152, Leningrad.

Feely, H. W. and Kulp, J. L. 1957. "Origin of Gulf Coast salt-dome sulfur deposits." Bull. Amer. Assoc. Petrol. Geologists, 41, No. 8, p. 1802.

Gasanov, M. V. 1961. "Biogenic reduction of sulfates in formations during flooding of oil deposits by sea water." Present volume, see p. 79.

Ginzburg-Karagicheva, T. P. 1926. "Microbiological investigations in the sulfur-saline waters of Apsheron." Azerb. neft. khoz-vo, No. 6-7.

Hunt, N. F. 1915. "The origin of the sulfur deposits of Sicily." Econ. Geol., 10, No. 6, p. 543.

Ingerson, E. 1954. Nonradiogenic Isotopes in Geology. Isotope Geology [Russian translation from English], IL, Moscow.

Isachenko, B. L. (1927). Microbiological Investigations on Mud Lakes [in Russian], Selected Works, 1, Izd. AN SSSR, 1951.

Isachenko, B. L. 1951. Microorganisms as Geological Factors [in Russian], Selected Works, 2.

Ivanov, M. V. 1956. "The use of isotopes in studying the intensity of sulfate reduction in Belovod' Lake." Mikrobiologiya, 25, No. 3.

Ivanov, M. V. 1957a. "The role of microorganisms in the formation of sulfur deposits in the hydrogen-sulfide springs of the Sergievsk mineral waters." Mikrobiologiya, 26, No. 3.

Ivanov, M. V. 1957b. "The participation of microorganisms in the formation of sulfur deposits at Shor-Su." Mikrobiologiya, 26, No. 5.

Ivanov, M. V. 1957c. Dissertation: The Role of Microorganisms in the Formation and Destruction of Deposits of Native Sulfur [in Russian], OBN, AN SSSR, Moscow.

Ivanov, M. V. 1958. The Use of Isotopes for Studying the Role of Microorganisms in the Formation of the Sulfur Deposit at Shor-Su [in Russian], Trudy Konf. po primeneniyu izotopov i izluchenii v nauke i narodn. khoz-ve, Moscow.

Ivanov, M. V. 1959. Role of Sulfate-reducing Bacteria in the Formation of Hydrogen Sulfide in Ground Water of Upper Tortonian Strata [in Russian], Mineralogical Collection, L'vovsk. geol. obsh., 1959, No. 13.

Ivanov, M. V., Lyalikova, N. I., and Kuznetsov, S. I. 1958. "The role of sulfur-oxidizing bacteria in the weathering of rocks and sulfide ores." Izv. AN SSSR, seriya biol. No. 2.

Ivanov, M. V. and Ryzhova, V. N. 1960. "The intensity of hydrogen-sulfide generation in some littoral sediments of the Barents Sea." Doklady Akad. Nauk, SSSR, 130, No. 1.

Ivanov, M. V. and Terebkova, L. S. 1959a. "A study of the microbiological processes of hydrogen-sulfide formation in Solenoe Lake, Communication 1." Mikrobiologiya, 28, No. 2.

Ivanov, M. V. and Terebkova, L. S. 1959b. "A study of the microbiological processes of hydrogen-sulfide formation in Solenoe Lake, Communication 2." Mikrobiologiya, 28, No. 3.

Jones, G. E. and Starkey, R. L. 1957. "Fractionation of stable isotopes of sulfur by microorganisms and their role in deposition of native sulfur." Appl. Microbiol., 5, No. 2, p. 111.

Jones, G. E., Starkey, R. L., Feely, H. W., and Kulp, J. L. 1956. "Biological origin of native sulfur in salt domes of Texas and Louisiana." Science, 123, No. 3208, p. 1124.

Karavaiko, G. I. 1960. "Microzonal distribution of oxidizing processes in the sulfur ore of Rozdol deposit." Mikrobiologiya, 29, No. 6.

Khudyakov, Ya. P. 1954. "Present status and objectives of microbiology of soils." Mikrobiologiya, 23, No. 3.

Kuznetsov, S. I. 1952. Role of Microorganisms in the Cycle of Substances in Lakes. Izd. AN SSSR, Moscow.

Kuznetsov, S. I. and Sokolova, G. A. 1960. "Some data on the physiology of Thiobacillus thioparus." Mikrobiologiya, 29, No. 6.

Kuznetsova, V. A., Ashirov, K. B., Gromovich, V. A., Ovchinnikova, I. V., and Kuznetsov, S. I. 1957. "An experiment on suppressing the growth of sulfur-reducing bacteria in the oil horizon of the Kalinovo deposit." Mikrobiologiya, 26, No. 3.

Lyalikova, N. N. 1957. "A study of the process of utilizing carbon dioxide by purple sulfur bacteria in Belovod' Lake." Mikrobiologiya, 26, No. 1.

Macnamara, J. and Thode, H. G. 1951. "The distribution of S^{34} in nature and the origin of native sulfur deposits." Research, 4, No. 12, p. 582.

Malyshek, V. T. and Gasanov, M. V. 1959. "A study of sulfate reduction in mixtures of marine and alkaline formational waters." Trudy AzNII po dobyche nefti, No. 8.

Malyshek, V. T. and Maliyants, A. A. 1935. "Sulfur bacteria in the formational 'rose' waters of the Surukhan deposit and their significance in the geochemistry of the waters." Doklady Akad. Nauk, SSSR, 3, No. 5.

Malyshek, V. T., Maliyants, A. A., and Reinfel'd, E. A. 1935. "The presence of sulfur bacteria in the formational rose waters of the Surukhan oil field and the geochemical significance of this factor." Azerb. neft. khoz-vo, 7-8.

Miller, L. 1949. Cited by Beerstecher, 1954.

Murzaev, P. M. 1939. "A brief survey of theory and some communications on the origin of bedded sulfur deposits." Trudy Voronezhsk. gos. un-ta, 11, geol.-pochv. otd., No. 3.

Randama, K. 1956. Isotope Geology [Russian translation from English], IL, Moscow.

Rubenchik, L. I. 1948. Microorganisms and Microbial processes in Saline Basins of the Ukraine SSR [in Russian], Izd. AN Ukr. SSR.

Schwartz, W. 1958. "Die Schwefelspezialisten unter den Microorganismen." Handbuch d. Pflanzenphysiol. 9, p. 89.

Shturm, L. D. 1937. "The study of microflora in sulfur-bearing rocks." Mikrobiologiya, 6, No. 4.

Shturm, L. D. and Simakova, T. L. 1928. "Microbiological studies of sulfur samples from Crimean and Turkestan deposits." Doklady Akad. Nauk, SSSR, No. 8.

Simakova, T. L. 1961. "Bacterial changes of oil and its components in anaerobic environments." Present volume, see p. 61.

Sokolov, A. S. 1958 a. Dissertation: Basic Pattern of Geological Structure and Distribution of Sedimentary Deposits of Native Sulfur [in Russian], MGU, geol. fak-t.

Sokolov, A. S. 1958b. "Basic pattern of geological structure and distribution of sedimentary deposits of native sulfur." Sov. geologiya, No. 5.

Sokolova, G. A. and Sorokin, Yu. I. 1957. "Bacterial reduction of sulfates in muds of the Rybinskoe Reservoir." Mikrobiologiya, 26, No. 2.

Sokolova, G. A. and Sorokin, Yu. I. 1958. "Determination of intensity of bacterial reduction of sulfates in the sediments of the Gorky Reservoir, by means of S^{35}." Doklady Akad. Nauk, SSSR, 118, No. 2.

Stutzer, O. 1911. "Über genetisch wichtige Auschlüsse in den Shwefelgruben Siziliens." Zschr. Deutsch. geol. Ges., No. 1, p. 8.

Subba Rao, M. S., Iya, K. K., and Sreenivasaya, M. 1949. "Microbiological formation of elemental sulfur in coastal areas." 4th Internat. Congr. Microbiol. Copenhagen, Rep. of Proceedings, p. 494.

Thode, H. G., Kleerekoper, H., and McElcheran, D. 1951. "Isotope fractionation in the bacterial reduction of sulfates." Research, 4, No. 12, p. 581.

Thode, H. G., Macnamara, J., and Fleming, W. H. 1954. "Sulfur isotope fractionation in nature and geologic and biological time scales." Isotope Geology [Russian translation], IL, Moscow.

Uklonskii, A. S. 1940. Paragenesis of Sulfur and Petroleum [in Russian], Izv. Usb. fil. AN SSSR, Tashkent.

Vernadskii, V. I. (1912-1922). An Experiment in Descriptive Mineralogy [in Russian], Selected Papers, 2, Izd. AN SSSR, 1955.

Vernadskii, V. I. 1927. History of the Minerals of the Earth's Crust [in Russian], 1, No. 2, Leningrad.

Vinogradskii, S. N. 1924. "The direct method in microbiological investigation of soils." Microbiology of Soils [in Russian], Izd. AN SSSR, 1952.

Vinogradskii, M. N. 1945. "Principles of ecological microbiology." Included in Microbiology of Soils [in Russian] Izd. AN SSSR, 1952.

ZoBell, C. E. 1947. Cited by Beerstecher, 1954.

SOME SYSTEMATIC RELATIONS IN THE DISTRIBUTION OF NATURAL GASES AND MICROORGANISMS IN THE ZONE OF OIL AND GAS DEPOSITS

G. A. Mogilevskii

(All-Union Scientific-Research Institute for Petroleum Geological Exploration)

In previously published works on grounds for geochemical methods of prospecting for oil and gas and on geological interpretation of the results of such methods, it has been pointed out that ground water in the upper sedimentary sequence contains a high concentration of methane and heavy hydrocarbons, and also that there is a bacterial population oxidizing these gases, within local zones, that spreads over oil and gas deposits (Mogilevskii, 1957, 1958).

Very detailed investigations, made in this field by a thematic group of VNIGNI (All-Union Scientific-Research Institute for Petroleum Geological Exploration) in 1957-59 in parts of western Bashkiria, the Orenburg district, and northwestern Ciscaucasia, have demonstrated in a number of new examples the close connection between the indicated characteristics and previously known and newly developed oil and gas fields in these regions.

During drilling of exploratory holes in the investigated regions, mud logs were prepared at the same time that studies were made on the gas and bacterial content of the ground water. These logs gave information on the nature of migration of the hydrocarbon gases from the gas-oil deposits toward the upper layers of the earth.

A comparatively uniform increase in hydrocarbon concentration with depth has been established for the regions of the northern Caucasus.

The migration process is more complex in the districts of western Bashkiria. Here one may observe a low concentration of hydrocarbon gases in parts of the section between zones of maximum concentrations in Carboniferous and Lower Permian strata. It may be assumed that the gases in the intervening sequences of rock migrate through microfractures and dislocations. Migration in these rocks was apparently most extensive during past geologic epochs; in this connection a definite correlation is observed between local commercial deposits in Carboniferous rocks and noncommercial appearances of oil and gas in Lower Permian strata.

The purpose of the present report is to sum up the work of using gas-biochemical methods of prospecting for oil and gas in western Bashkiria, the Orenburg district, and the northern Caucasus and also to discuss the results of investigating the effect of hydrocarbon-oxidizing microorganisms on the diffusion of combustible gases through water and water-bearing sand.

Recent experiments make it possible, in some measure, to interpret anomalies in the distribution of hydrocarbon gases in the strata overlying deposits of oil and gas.

Western Bashkiria

In western Bashkiria and the southeastern part of the Orenburg Oblast studies were made on the ground water in Permian strata at several localities that are distinguished by variations in the character of the underlying gas- and oil-bearing strata of Carboniferous and Devonian age.

Among the investigated areas there are oil deposits that have long been exploited (Tuimazy, Serafimovskoe, Shkapovskoe) and some recently discovered as well as areas that were explored after the gas-biochemical survey (Ik-Baza, Bakaly, and others). The testing of the gas-biochemical survey data in the Bakaly area, north of Tuimazy, is of interest because of the methods employed (Fig. 1).

The Drilling Trust of the Tuimazy Petroleum Industry has drilled several exploratory holes. Two of these were situated in zone of a gas-biochemical anomaly. Hole 21, situated in the center of a gas and bacterial anomaly for methane in the Bakaly area, yielded only noncommercial shows of oil and gas in Permian rocks. Hole 27, to the southeast, cut an oil show in the Yasnaya Polyana subseries C_3. Tests indicate the yield of this well to be about 5 tons per day. Both holes are within anomalous zones, distinguished by the presence of propane-oxidizing bacteria. Another example of successful testing of the gas-biochemical method is found in the results of exploratory work in the Serafimovskoe district (Fig. 2).

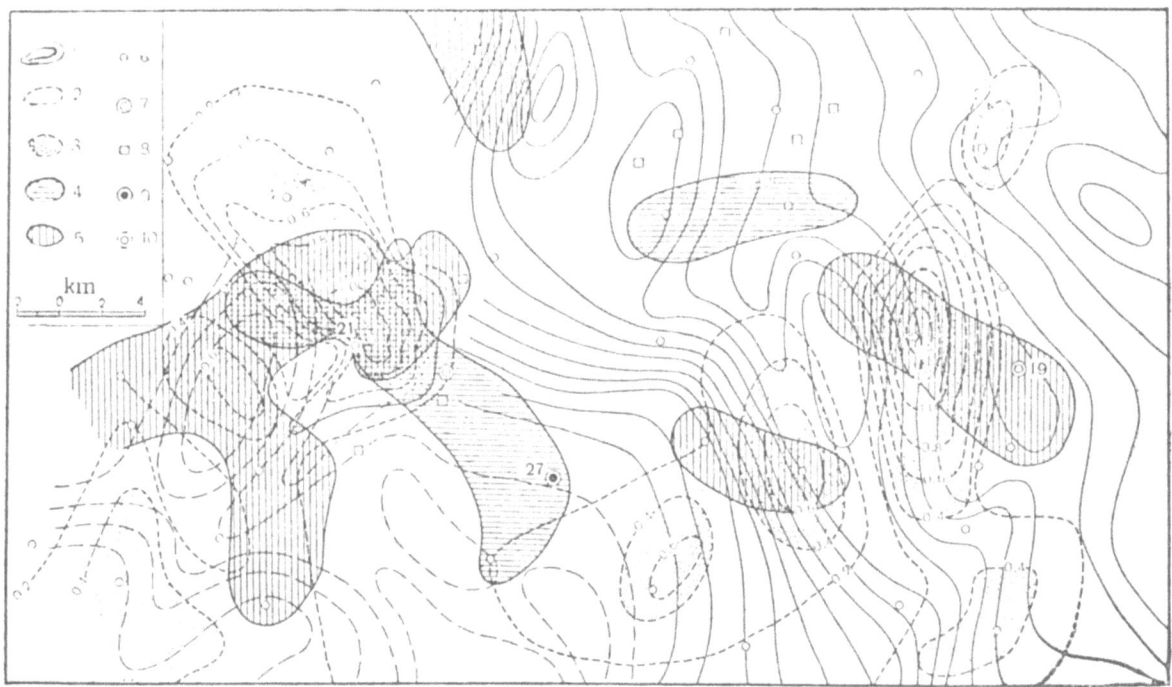

Fig. 1. Results of a gas-biochemical survey of the water, in 1958-59, in the Bakaly region of western Bashkiria. Prepared by M. M. Gutman. 1) Structure contours on top of the Artinskian strata; 2) isopleths for equal content of hydrocarbon gases, cm^3/liter; 3) zones with a content of hydrocarbon cases greater than 0.4 cm^3/liter. Zones of distribution: 4) propane-oxidizing bacteria; 5) methane-oxidizing bacteria. Investigated sites: 6) springs; 7) boreholes; 8) wells; 9) holes with oil shows; 10) abandoned holes.

Here, to the south of the Serafimovskoe oil field, drilling has revealed a new productive area. It was found to lie in the zone of a gas-biochemical anomaly that came to light during reconnaissance work in 1957. Flowing oil was obtained from two holes, from Upper Devonian rocks, one within the zone of the anomaly and the other situated immediately next the anomaly.

Furthermore, the results of a water survey and of gas measurements in the holes have been partially confirmed by further exploratory work in the northern part of the Orenburg district, in the region of the Efremovo-Zykovskii multilayered field, and also near the Izmailovo gas deposit.

In comparing the results of water surveys for different indices, one may note a clearer differentiation in the relationships of the propane-oxidizing bacteria (Fig. 3).

At the second site hydrocarbon gases and methane-oxidizing bacteria are found.

A comparison of the gas and oil contents in the investigated areas with the results of gas-biochemical studies of the ground water in the upper zone of sedimentary rock shows that propane-oxidixing bacteria grow most abundantly in those areas where the gas and oil of a deposit occur at several levels or where the gas factor of an oil deposit is great.

A comparison of the gas-biochemical survey of the water with a semi-automatic mud log, summed up for the Upper and Lower Permian rocks and the Carboniferous strata, indicates that the gas and oil beds nearest the surface,

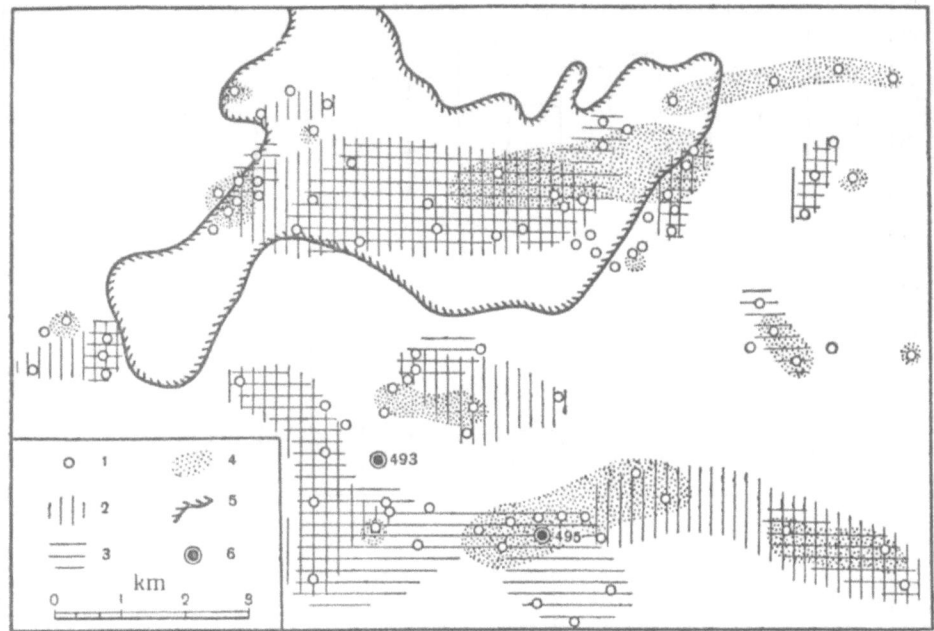

Fig. 2. Distribution of methane- and propane-oxidizing bacteria in dissolved hydrocarbon gases in the ground water of Upper Permian strata in the districts of the Serafimovskoe, Leonidovskoe, and Konstantinovka oil deposits from data of the 1957 survey. Prepared by Yu. N. Lisitsyna. 1) Localities of water samples. Zones of distribution in the ground water of the Upper Permian strata; 2) propane-oxidizing bacteria; 3) methane-oxidizing bacteria; 4) dissolved hydrocarbon gases; 5) outline of oil-bearing zone in bed D_1; 6) holes that in 1959 opened up a new oil deposit in the Devonian.

characterized by the development of a gas-biochemical anomaly in the ground water of the Upper Permian strata, are noncommercial deposits of oil and gas in the Kungurian and Artinskian series. There are grounds for assuming a genetic connection between these deposits and the productive horizons in the Carboniferous.

Search was also made for ethane- and butane-oxidizing bacteria in a large number of water samples from western Bashkiria during the ordinary determinations of methane- and propane-oxidizing bacteria.

As a result of these studies (data of Z. S. Smirnova), it has been ascertained that butane- and ethane-oxidizing bacteria have a rather restricted distribution in the ground water of the Upper Permian strata of western Bashkiria (see table). However, almost all discoveries of these organisms were made in already known oil deposits or in deposits opened up since the water survey.

Methane-oxidizing bacteria were found in the greatest number of samples; propane-oxidizing organisms come next in abundance. Ethane and butane-oxidizing microflora are least abundant.

Methane- and propane oxidizing organisms were studied chiefly in the deeper aquifers of western Bashkiria, in Carboniferous and Devonian strata.

It was found in this study that propane-oxidizing bacteria decreased more rapidly with depth than methane-oxidizing forms.

For individual areas investigated in the Bakaly and Mancharovskoe regions, studies were made on the effect of geotectonic factors on the development of geochemical anomalies identified in the ground water of Upper Permian strata.

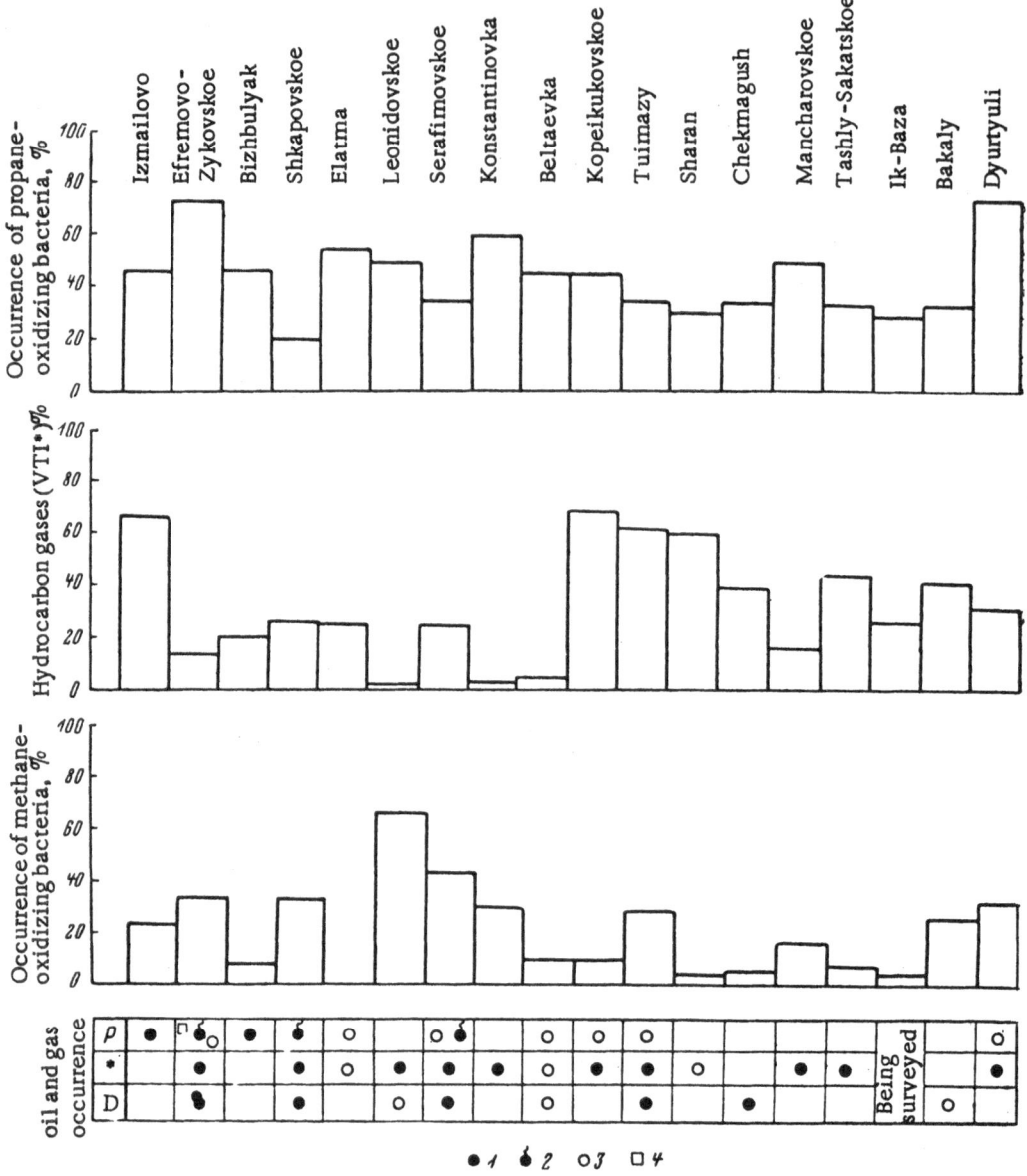

Fig. 3. Average gas-biochemical indices and occurrence of bacteria in aquifers of Upper Permian strata for deposits and exploratory areas in western Bashkiria and the northeastern part of the Orenburg Oblast, compared with the gas and oil content of the productive horizons. 1) Oil show; 2) gas show; 3) oil absent; 4) bed flooded.

Correlation graphs compared for this purpose have shown that propane-oxidizing bacteria and, in part, hydrocarbon gases are controlled in great measure by the structural plan of the Artinskian series, where noncommercial accumulations of oil and gas are present, rather than by the Spirifer horizon (which is nearer the surface), within which a number of gently inclined structures are found not coinciding with the local uplifts in the Artinskian series or the deeper Carboniferous strata.

One's attention is drawn to the individual zonal development of methane- and propane-oxidizing bacteria in the individual anomalies; this is noted particularly in the Bakaly exploratory district and in the Serafimovskoe oil field. As exploratory work has shown, independent zones of propane-oxidizing bacteria are of greatest practical significance in connection with the presence of oil in the ground.

---

* According to the All-Union Heat-Engineering Institute.

TABLE 1. Distribution of Methane-, Ethane-, Propane-, and Butane-Oxidizing Bacteria in Ground Waters of Western Bashkiria and the Azov-Kuban Basin (abundance in conventional units)

Region of investigation	Total No. of samples	Distribution of bacteria oxidizing the following							
		methane		ethane		propane		butane	
		% detected	average abundance	% detected	average abundance	% detected	average abundance	% detected	average abundance
Western Bashkiria	147	65.5	46.0	9	2.26	26.0	11.0	7.4	1.15
Azov-Kuban basin	61	58.98	39.5	6.65	3.6	45.3	27.5	18.7	8.5
Central Asia	20	26	—	18.1	—	26	—	20.8	—

Zones of less abundant methane- and propane-oxidizing bacteria are associated with increased carbon-dioxide content in many random segments of the structures.

Northwestern Ciscaucasia

In this region a gas-biochemical survey was made of the northeastern border of the Azov-Kuban basin, within which a number of gas-bearing structures have been discovered in recent years. Studies were made on ground water in Quaternary and Upper Pliocene deposits.

In considering the abundance of propane- and methane-oxidizing bacteria, and also the average content of dissolved hydrocarbon gases, in the ground water of the Azov-Kuban basin (Fig. 4), one may note that highest values were obtained in areas where known gas deposits are present at depth: Kushchevskaya, Leningradskaya, Statominskaya, and Kanevskaya. High values were also obtained for the Shkurinskaya uplift, which has not yet been explored (data of A. N. Sukhovoya and V. B. Neiman).

Along with the high values for areas corresponding to buried gas-bearing structures, very small quantities of index microflora and hydrocarbon gases have been noted for the Eisk district and the Novoshcherbinovskaya area, where commercial deposits of gas have not yet been established.

Thus, the general character of the distribution of hydrocarbon gases and microorganisms that oxidize methane and its homologues points to a fundamental relationship between the developed anomalies and the distribution of oil and gas deposits at depth.

An analysis of the survey results indicates that carbon dioxide and hydrogen sulfide (acidic gases) exert the strongest influence on the distribution of microorganisms. In places where these gases are most abundant a considerable decrease is noted in the number of hydrocarbon microflora and a corresponding increase in the concentration of dissolved methane in the water.

The results of the gas-biochemical study of the ground water of the Azov-Kuban basin (Fig. 5) are of value as support for some of the theoretical aspects of the method, since they demonstrate the possibility of discerning traces of migration within the upper zone of sedimentary rocks when the buried gas-oil deposits are covered by a thick sequence of homoclinal strata.

Consequently, under such geostructural conditions, the vertical migration of hydrocarbons occurs simultaneously with movements of these substances along bedding planes.

Along with the gas-biochemical studies in regions of gas and oil deposits, the same group of studies were made on the area of the Kaluga uplift, known to contain no deposits of oil and gas. Neither propane-oxidizing bacteria or heavy hydrocarbons were detected in the ground water of this region. Methane-oxidizing bacteria were found only in small numbers.

Apart from questions associated with the application of the method under actual geologic conditions, the investigations in the regions of western Bashkiria and northern Ciscaucasia have brought to light other systematic relationships, which are of value for proper interpretation of the geochemical investigations (Data of the Geochemical Conference, 1960).

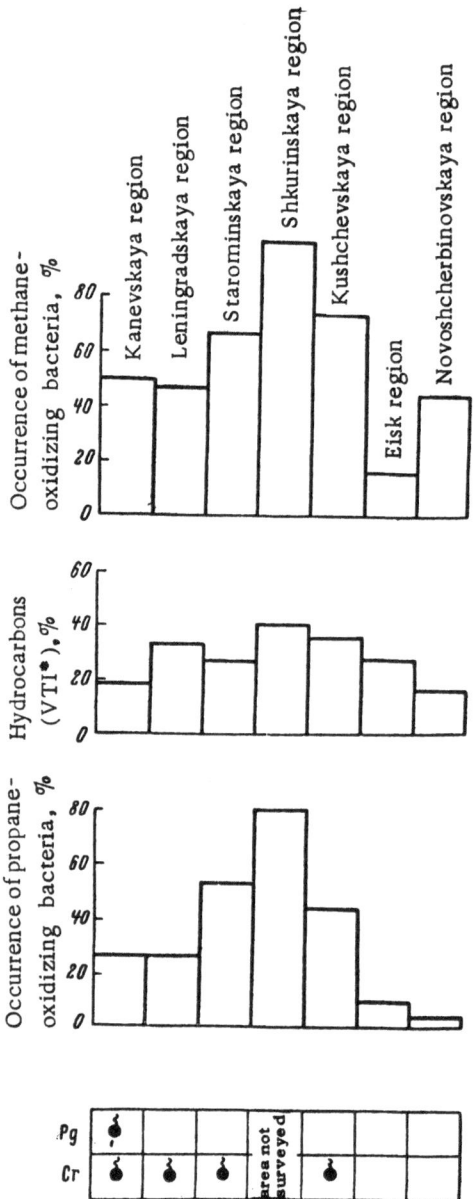

Fig. 4. Average values of gas-biochemical indices and occurrence of bacteria for deposits and exploratory areas in the Azov-Kuban basin, compared with the presence of commercial gas in Paleogene (Pg) and Cretaceous (Cr) rocks. Symbols are the same as in Fig. 3. *) According to the All-Union Heat-Engineering Institute.

Experiments to Illuminate the Effect of Microorganisms on the Diffusion of Gases through Water and Water-Bearing Sand

During the gas-biochemical survey of the water, a number of questions arose concerning the behavior of hydrocarbon-oxidizing microorganisms in ground water in connection with the diffusion of gases through the aqueous medium.

The thought was to establish special model experiments that would reproduce natural conditions in so far as this was possible without complicated laboratory equipment. The experimental work was done by V. M. Bogdanova and B. S. Cherkinskaya.

In setting up the experiments, the following objectives were kept in mind.

1. Ascertaining the role of hydrocarbon-oxidizing microorganisms during the diffusion of gases through an aqueous medium and through water-bearing sands.

2. An explanation of the precise proportions in the system air—water—gas that will be most favorable for the growth of microorganisms: where there is a higher partial pressure of hydrocarbons? or of oxygen?

3. A solution to the problem of whether a sequence of sand separating hydrocarbon gases from microorganisms will prevent the latter from consuming the gases.

Propane-oxidizing bacteria were used in the experiments; these organisms were introduced into the experimental vessel, but were absent in the control. Propane and air were introduced in all the experiments in the same quantities, namely 100 cm^3 of propane and 170 cm^3 of air. The experiments were carried out with specially constructed apparatus (Fig. 6).

The layer of sand was 60 cm thick. In one experimental vessel a culture of propane-oxidizing bacteria was introduced into the upper arm (Expt. 1), so that the layer of sand separated the bacteria from the propane. In another experimental vessel, propane-oxidizing bacteria were placed in the lower arm (Expt. 2). Here the sand separated the bacteria from the oxygen air. The control vessel contained no bacteria. The vessels were incubated in a thermostatically controlled chamber at 32° for 45 days, after which bacterial and gas analyses were made (see Table 2).

Despite the preliminary aspect of the experiments, the results clearly show that the natural process of penetration of gaseous hydrocarbons through water and water-bearing sand may be implemented by the action of hydrocarbon-oxidizing bacteria when these bacteria grow at a certain distance from the source of the hydrocarbons.

Fig. 5. Sketch of the northern border of the Azov-Kuban basin on the base of the Santonian-Senonian, showing the principal data of the gas-biochemical study of the water, made in 1959, on distribution of gases and microflora. Prepared by V. B. Neiman. 1) Zones of methane-oxidizing bacteria, having an abundance greater than 100 conventional units; 2) zones of propane-oxidizing bacteria, having an abundance greater than 100 conventional units; 3) zones containing more than 0.05% methane (from data of an UKhTG instrument); 4) zones containing more than 0.02% heavy hydrocarbons; 5) structure contours on base of Santonian-Senonian; 6) isopleths of equal content of hydrocarbon gases, cm^3/ liter (from data of a VTI instrument).

TABLE 2. Analysis of Gas in Volume Percentage for Experimental and Control Vessels

Gas	Exptl. vessel 1		Exptl. vessel 2		Control vessel	
	O_2	C_3H_8	O_2	C_3H_8	O_2	C_3H_8
Free gas from the upper arm	18.6	0	11.8	2.7	20.2	0.74
Free gas from the lower arm	6.1	82.5	0.55	80.5	0	81.4
Dissolved gas from the upper layer of liquid	15.5	0	20.7	0.7	20.6	0
Dissolved gas from the lower layer of liquid	8.4	15.4	18.4	4.3	15.0	6.8

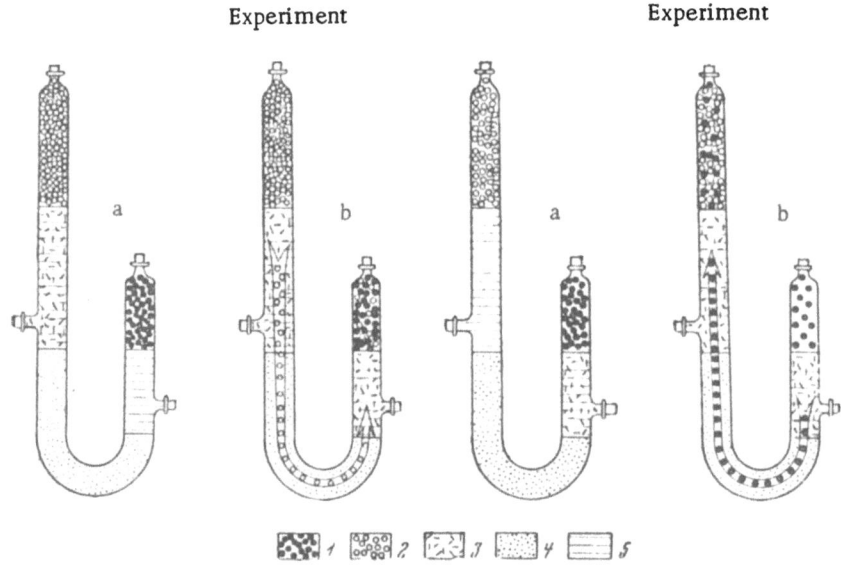

Fig. 6. Diagrammatic sketch of experiments to investigate influence of microorganisms on the diffusion of gases; period of incubation, 45 days; temperature, 32°. a) At beginning; b) at end; 1) propane; 2) air; 3) propane-oxidizing bacteria; 4) and 5) nutrient medium.

When the hydrocarbon-oxidizing microflora grow in direct contact with hydrocarbon gas, on the other hand, the bacteria prevent diffusion of the gas, playing the role of a bacterial filter (Expt. 1).

On the basis of these model experiments, it may be suggested that under natural conditions bacteria that oxidize hydrocarbon gases will increase the migration of these gases from lower-lying beds toward the earth's surface.

Conclusions

1. Investigations have been made in a number of places in the USSR on the distribution of hydrocarbon-oxidizing bacteria in subsoil horizons and waters for the purpose of exploring for oil and gas deposits. Definite patterns were discovered in the distribution of natural gases and microorganisms in the zone of such deposits. Positive results of such exploration have been confirmed, in most cases, by drilling.

2. A distinct differentiation in gas saturation is noted in the section for propane-oxidizing bacteria. The greatest concentration of propane-oxidizing bacteria in surface horizons is found in areas of multilayered deposits of gas and oil. The number of propane-oxidizing bacteria decreases more rapidly with depth than methane-oxidizing bacteria.

3. An inverse relationship is observed between the content of carbon dioxide in subsoil gases and the number of hydrocarbon-oxidizing bacteria.

4. Earlier observations on the absence of propane-oxidizing bacteria in regions known to contain no oil have been confirmed.

5. A comparative study of the different species of index bacteria has shown that methane-oxidizing bacteria are found in the greatest number of samples; these are followed, in decreasing order, by propane-, ethane-, and butane-oxidizing bacteria.

6. Model experiments show the possibility of diffusion of hydrocarbon gases through water-bearing layers. The diffusion of the gas may be increased or diminished, depending on the nature of growth of the hydrocarbon-oxidizing bacteria.

LITERATURE CITED

Geochemical Methods of Prospecting for Oil and Gas [in Russian], Data of the Geochemical Conference that Convened in Moscow in 1958, Presidium of the Academy of Sciences, USSR, under the editorship of V. A. Sokolov, 1960.

Mogilevskii, G. A. 1957. "A study of the nature of gas-bacterial anomalies from drilling data and mud logs." Twentieth International Geological Congress, Proceedings of Soviet Geologists, 1, Gostoptekhizdat.

Mogilevskii, G. A. 1958. "Experiments on gas-biochemical surveys of water for prospecting for oil and gas deposits." Sov. geologiya, No. 11.

MICROFLORA IN GROUND WATER AND ITS SIGNIFICANCE
IN THE FORMATION OF NATURAL GASES, OIL, SULFUR
AND SOME GASEOUS COMPONENTS

M. E. Al'tovskii

(All-Union Scientific-Research Institute of Hydrogeology and Engineering Geology, Moscow)

In order to solve a number of problems on theoretical hydrogeology and, above all, in order to study the constructive action of ground water as a geologic agent, the All-Union Scientific-Research Institute of Hydrogeology and Engineering Geology began in 1954 broad regional investigations on the distribution and abundance of the various physiological groups of bacteria in ground water. At the same time, quantitative and qualitative studies were made on the composition of the inorganic and organic substances, gases, different microcomponents, concentration of hydrogen ions, and oxidation-reduction potential. This work was carried out by the hydrogeologist V. M. Shvets and a group of organic chemists (Bykova, Brodovskaya, Dudova, and Frantskevich) under the direction of the author of the present paper. Field and laboratory microbiological studies in all the regions enumerated below were made by Z. I. Kusnetsova.

The investigations embrace, in various degrees of thoroughness, ground water in the Terek-Dagestan oil district, some aquifers in the Terek-Kuma (Apsheron horizon) and Moscow artesian basins, and ground water in the oil fields eastern Georgia, Cheleken, and western Turkmenia.

It is well known that microbiological life in the shell of sedimentary rocks below the water table of the first aquifer below the earth's surface is possible only in ground water.

Numerous sanitary-hygienic investigations long ago established the fact that beds impermeable to water, even when relatively thin, preserve shallow ground water, normally used for household-drinking purposes, rather free from bacterial contamination. Therefore, the principal and most active replenishment of microflora in ground water occurs through exposed and covered recharge zones. Through these "hydrogeological windows" organic material, serving as food for many kinds of microorganisms, is also introduced from the earth's surface.

If we have in mind only the sequence of sedimentary rocks, then the first recharge zone, or "hydrogeological window," appeared on the earth in Cambrian time. After this, with increasing thickness of the sedimentary cover, the number of recharge zones grew rapidly, while, it is true, the area about each one, leading to it, simultaneously grew smaller. Thus, microflora began to inhabit ground water at least as early as the Cambrian and have continued to inhabit it to the present time.

Along with underground flow, moving down the slope of beds to the depths of the sedimentary shell, microorganisms also spread out; in individual zones of these underground streams, which extend as far as several hundred kilometers and reach depths of several kilometers, conditions exist that are favorable for the growth of some physiological groups of microorganisms and that are unfavorable for other groups.

The deepest segments of aquifers, containing ground water with mineralization that sometimes reaches 250-500 g of salt per liter, and having a temperature up to 80-90° and even as high as 150-180°, must serve as sites of destruction of microorganisms and, correspondingly, as sites of accumulation of tremendous quantities of biomass over long periods of geologic time. These zones must also be the sites of accumulation of some of the products forming through the life activity of microorganisms in the relatively higher parts of the aquifers, products that move downward to the lower parts of the sedimentary shell by ground-water currents.

In the field work samples of water for microbiological studies and for other analytical work were collected (depending on the geology of the particular region and on the presence of drill holes) so that a representative sampling was obtained for the microflora of the aquifers throughout the entire extent of the beds, from the recharge zones to the discharge zones or to the deepest segments of the beds.

Water samples were collected from springs, very rarely from dug water wells (sanitarily safe), and from drill holes (flowing or in continuous operation, or after prolonged pumping when the entire pipe is filled with water to the height at which the sample is taken).

The distribution and abundance of some microorganisms, as should be expected, are closely connected with the organic material in the ground water.

In all the 302 water samples that were analyzed, organic material was found without exception in some amount. Expressed in the content of organic carbon, this material ranged from 4 to 90.8 mg/liter. Organic material is found everywhere in ground water, both in oil districts and in non-oil districts, and also in water of the shallowest circulation (descending springs in the recharge zone of aquifers) and in the deepest water. The greatest depth from which a water sample was taken was 3200 m (a flowing well in Azek-Suat). A rather large amount of organic material (averaging 9 mg/liter) is contained in rising springs in the discharge zone of ground water in the Grozny region.

The total quantity of organic material contained in ground water is tremendous. According to the computations of Shvets, approximately 24,000 tons of organic material is introduced annually into the Carboniferous waters of the Moscow artesian basin.

Let us make one more computation to characterize the total amount of organic material passing through just the Karagan and Chokrak strata of the Grozny region. It is known that before the oil deposits in this region were first exploited the total yield of all rising springs in the discharge zone amounted to 18,200 m^3/day. These springs contain on the average about 10 mg/liter organic material. From this it follows that for one-fourth the period of transit to the surface (about 500,000 years), approximately 32,500,000 tons of organic material was discharged.

The organic material in the ground water of the Makhachkala district consists of dissociated and undissociated compounds. In the acetone fraction, the organic material consists of low-molecule compounds that pass almost completely through semipermeable membranes. They embrace the following functional groups: hydroxyl, carbonyl, ether, and carbon atoms with double bond. The organic material includes carbohydrates—sugar and uronic acid (hundredths of mg/liter); free amino acids, glycine, lysine, asparagine, α-alanine, and glutamine acid (hundredths of mg/liter); crenic and apocrenic acids; phenols (hundredths of mg/liter); purine base and pyrimidine base (thousandths of mg/liter); aromatic substances such as thymol (thousandths of mg/liter); terpenes; hydrocarbons of the solar fraction (thousandths of mg/liter); porphyrins; naphthenic acids; resin acids and substances forming after hydrolysis of tar-like products (a few milligrams per liter); and fatty acids.

It should be noted that, on the average, organic substances and naphthenic acids are by no means less abundant in ground (upper), stream, and swamp waters of European USSR, but are even sometimes more abundant, than in deeper ground waters in the investigated zones of oil and gas deposits. Weakly mineralized waters of oil regions (eastern Georgia) also have high contents of organic carbon (from 9.7 to 45.4 mg/liter) and naphthenic acids (from 120 to 1420 mg/liter). This complex group of organic substances, migrating with the ground water, undoubtedly undergoes some sort of profound change as it moves down the dip of the water-bearing beds, since the ratio of organic carbon to organic nitrogen changes from 10.5 to 32.4; i.e., the organic material appears to become gradually enriched in carbon and impoverished in nitrogen.

The total amount of organic material increases approximately sixfold down the dip of the water-bearing beds from the recharge zone to oil deposits, and approximately twofold from the recharge zone to the discharge zone.

Ground water definitely has a completely independent source of supply of organic material, not associated with the rocks or with the destruction of oil deposits. This source is found chiefly in the remains of terrestrial plants, a fact supported by the presence in the organic material in ground water of such compounds as sugar, free amino acids, and pyrimidine bases.

The organic material in ground water includes remains of terrestrial plants in various stages of decomposition and transformation, and containing carbohydrates and products split off from them as well as albumins and products split off from them.

The presence of albumins is indicated by the intense development of such physiological groups of bacteria as the putrefactive bacteria and saprophytes; ground water is found to contain purine and pyrimidine bases (aldehydes and ketones) and compounds with the functional group OH, aromatic substances, and organic material profoundly altered by soil-making processes (such as crenic and apocrenic acids).

The profound physicochemical changes undergone by the organic material in ground water are, at the same time, accompanied by and complicated by the biochemical activity of microflora. The ground waters in the investigated region were found to contain the following physiological groups of bacteria: putrefactive, saprophytic, desulfurizing, denitrifying, cellulose-decomposing, sulfur-oxidizing, hydrogen-oxidizing, phenol-oxidizing, naphthalene-oxidizing, heptane-oxidizing, methane-oxidizing, and methane-forming in various environments. The total quantity of bacteria sometimes reaches several millions per milliliter, the number of viable bacterial cells almost always amounting to 94-100% of the total.

In the recharge zones of the water-bearing horizons of the Grozny-Dagestan region, putrefactive, saprophytic, phenol-oxidizing, denitrifying, cellulose, and methane-forming bacteria are abundant; hydrogen-oxidizing bacteria are rather uncommon, and desulfurizing and methane-oxidizing forms are even less abundant.

Somewhat different physiological groups of bacteria are more abundant in the discharge zone of these same horizons, particularly desulfurizing, sulfur-oxidizing, cellulose, and methane-forming, in an environment containing hydrogen and carbon dioxide; saprophytic and hydrogen-oxidizing bacteria are much less abundant, and phenol-, heptane-, and methane-oxidizing bacteria are extremely scarce.

The following physiological groups are most abundant in the ground water in oil districts of the Grozny region: bacteria that oxidize phenol, naphthalene, and heptane; methane forming bacteria are somewhat less abundant; such bacteria as desulfurizing, sulfur-oxidizing, denitrifying, and cellulose bacteria occur in about the same proportion as in the discharge zone; putrefactive, saprophytic, hydrogen-oxidizing, and methane-oxidizing bacteria are rather scarce.

The high mineralization in the ground water at the gas fields of the Dagestan region apparently depresses the activity of many physiological groups of bacteria; because of this, most of the groups mentioned above are completely absent or are very scarce, and only bacteria that cause carbon-dioxide, methane, and nitrogen fermentation continue to grow extensively.

In the highly mineralized waters (150-200 g/liter) of western Turkmenia, microorganisms are almost absent, although this statement is subject to still further verification.

In the ground water of the oil fields of eastern Georgia, the most abundant bacteria are denitrifying and cellulose-destroying forms; putrefactive bacteria are also rather abundant in the recharge zone.

Denitrifying and cellulose-destroying bacteria are widely developed in the petroliferous waters of the Terek-Kuma and Moscow artesian basins; methane-oxidizing and methane-forming bacteria are present in approximately equal numbers.

A composite study of the organic material and the microflora of ground waters has shown that the distribution and abundance of some physiological groups of bacteria depend closely on the quantitative and composition of the organic material migrating with the ground waters. The widely known process of desulfurization becomes intense only when SO_4^- ions in the ground water are accompanied by organic material.

Cellulose-destroying bacteria are abundant and widespread in the ground waters of all the investigated regions and in all parts of the water-bearing horizons (from recharge zones to discharge zones); this fact may be explained only by assuming that cellulose penetrates into the deep parts of the sedimentary shell together with the ground water.

The abundance of such bacteria as putrefactive and cellulose-destroying forms fluctuates through the years. For example, in the water of the Uitash Spring, which emerges in the discharge zone of Karagan waters in the Makhachkala region, the abundance of putrefactive and cellulose-destroying bacteria was determined as 5 on the scale (according to studies by Z. I. Kuznetsova); in 1956 the value was near zero, but the number of colonies obtained by inoculation on glucose-peptone agar were 2622 and 706 per milliliter, respectively.

The geological activity of microflora in recharge zones and in the upper parts of aquifers (at least to the depth of oxygen penetration) is manifested primarily in aerobic decomposition of the remains of terrestrial plants (cellulose, carbohydrates, albumins). Further decomposition in the deeper parts of the water-bearing strata occurs in a reducing environment and, for the most part, under anaerobic conditions.

The principal result of biochemical activity of microflora in ground water is the formation of such gases as carbon dioxide, methane, hydrogen sulfide, nitrogen, and, possibly, hydrogen, which substantially alter the composition of the ground water and lead to the formation of carbon-dioxide and hydrogen-sulfide mineral waters that may of balneological value, to the generation of the chief components in deposits of gas fuels, or to the production of the gaseous components of oil. Through the formation of hydrogen the necessary conditions may be produced for hydrogenation, leading to the formation of liquid hydrocarbons. The over-all scheme of biochemical activity of microflora and of the formation of the disseminated components of oil is shown in Fig. 1.

The basic material for the formation of carbon dioxide, methane, and, possibly, hydrogen is cellulose; down to the depth of oxygen penetration (where the downward velocity of the ground water is considerable this depth may be several hundred meters) carbon dioxide is chiefly formed, but at greater depths, or, more precisely, in a reducing environment, methane is formed. In intermediate zones of the water-bearing horizons both carbon dioxide and methane may form simultaneously.

Where methane-oxidizing bacteria are less abundant than methane-forming bacteria, methane accumulates directly in the aquifers over long periods of geologic time (Grozny-Dagestan district). In the opposite situation, the methane that is formed is almost completely consumed by methane-oxidizing bacteria (Carboniferous waters of the Moscow artesian basin).

Denitrifying bacteria in ground water at the recharge zone and in the upper parts of the water-bearing strata most likely live together with nitrifying bacteria. From data on the distribution of nitrogenous compounds in ground water, it should be stated that nitrification processes predominate in the upper parts of the water-bearing strata. However, as the ground water passes into the zone of reducing environment the nitrates begin to be reduced to free nitrogen. In this connection, a zone of ground water enriched in biogenic nitrogen forms in parts of the water-bearing strata, in the zone between carbon-dioxide and methane fermentation; because of this one may frequently find a zone of nitrogen-methane gas.

It has now been firmly established in hydrogeology that the mineralization of ground water increases uninterruptedly from the recharge zone down the dip of water-bearing strata, and bicarbonate and fresh waters sometimes are changed to typical sulfate waters. In these later waters, when organic material is present, desulfurication becomes a dominant process, leading to extensive formation of hydrogen sulfide. The loss of the sulfate ion from ground water limits the zone of hydrogen-sulfide fermentation downward along the flow of the ground water. The enrichment of the ground water in hydrogen sulfide in one geologic environment increases the amount of so-called acidic cases; in another it leads to the formation of hydrogen-sulfide mineralized waters of balneological value; and in yet a third it leads to the formation of some sulfur deposits.

For sulfur deposits to form it is necessary, above all, for large quantities of sulfate ions to accumulate in the ground water; this accumulation may be due to the solution of gypsum, to the decomposition of pyrite, or, chiefly, to gradual concentration over long periods of geologic time, thus creating regional zones of sulfate waters. Organic material also accumulates in these ground waters. The simultaneous presence of these two conditions creates an environment for intense hydrogen-sulfide fermentation. The formation of free sulfur from hydrogen sulfide and from hydrous sulfides in the ground water may be, in places, purely biochemical, taking place during oxidation by sulfur bacteria and sulfur-oxidizing bacteria with the secondary formation of sulfuric acid, or it may occur, partially, by oxidation of these initial substances by oxygen of the air during crushing and approach of the crest of the fold to the surface of the earth.

It has long been known that deposits of rock salt, sulfur, oil, and disseminated hydrocarbons occur in some sort of close paragenetic relationship. It is most likely that the paragenesis of these mineral deposits involves a common natural environment of formation. And this common environment entails required substances in the ground water and physicochemical and biochemical processes effective in ground water of intermediate and deep circulation in order for the deposits to be formed. The initial substances for the formation of salt deposits are mineral salts, for sulfur deposits hydrogen-sulfide waters, and for oil and natural gases the organic material of the ground water.

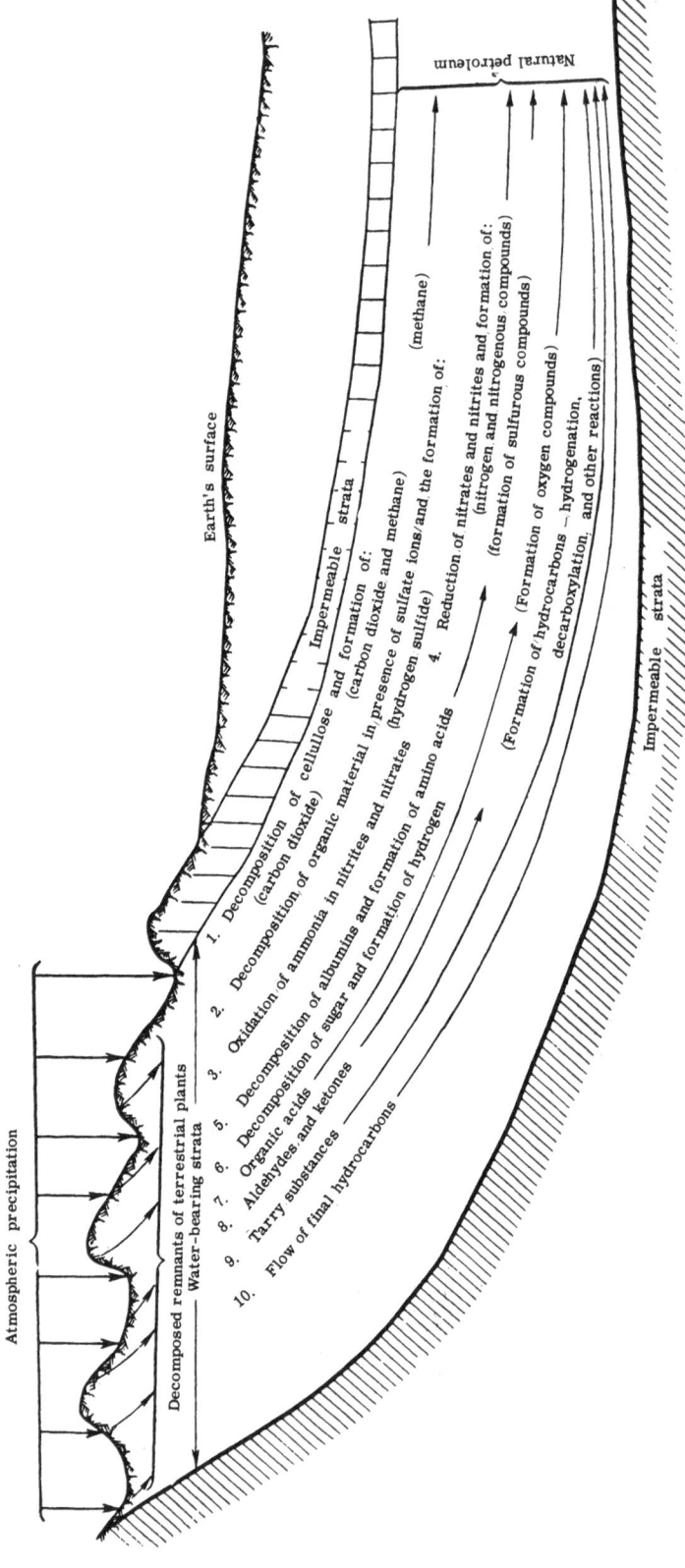

Fig. 1. General scheme of the biochemical activity of microflora and of the formation of the disseminated components of oil in homoclinally dipping water-bearing strata; arrows indicate direction of migration of organic material; the inscriptions concerning the generative processes and the products formed (enclosed in parentheses) indicate the approximate places of origin in the separate parts of the water-bearing strata.

Conclusions

1. Microflora in ground water began to spread through the recharge zone in the Cambrian and they have continued this process to the present time.

2. The group of dissolved organic substances alters qualitatively as the water moves along the strike of the bed, and the total quantity of organic material increases.

3. Ground water is enriched in organic material by the remains of buried terrestrial plants.

4. The microflora in a formation change in a direction away from the recharge zone toward the discharge zone; this change is associated with the character of the dissolved organic material and with the kind of salt in the water.

5. A scheme is presented to illustrate the processes by which the organic material dissolved in formational water is altered and changed to petroleum.

THE DISTRIBUTION OF MICROORGANISMS IN PRESENT-DAY AND ANCIENT CLAY-SILT SEDIMENTS

V. L. Mekhtieva

(All-Union Scientific-Research Geological-Prospecting Petroleum Institute, Moscow)

The study of the distribution of microorganisms in rocks of the lithosphere is very important for an explanation of the role of these organisms in geochemical processes at the various stages of diagenesis of rocks. In regard to the origin of oil there is special interest in the investigation of microflora in present-day and ancient clay and silt saltwater sediments, since most investigators associate the formation of primary oil with precisely such rocks.

The works of a number of scientists have established the presence of living microflora in present-day and ancient marine sediments (ZoBell, 1946; Beerstecher, 1954; Mekhtieva and Malkova, 1958; Kriss, 1959).

In deposits of the present-day Caspian and in the upper part of the Khvalisy beds (Upper Quaternary), living microflora have been traced down to a depth of 22.5 m below the surface of the floor (Table 1); a decrease in microbial inhabitants was noted with depth, both quantitatively and qualitatively. A rapid decrease in moisture has also been noted with increasing depth in the Caspian sediments, whereas the content of organic carbon decreases insignificantly and the amount of nitrogen remains practically the same. Sample 6, in which some resurgence of bacterial life was observed, was distinguished from other samples by a high content of moisture, carbon, and nitrogen, and by a low oxidation-reduction potential (Fig. 1)

TABLE 1. Physiological Groups of Bacteria in the Section of Caspian Sediments

Sample no.	Mud	Interval of column, m	Bacteria									
			ammon-ifying	denitrify-ing	glucose-ferment-ing	sulfate-reducing, on Van Delden medium with		cellulose		chitin de-stroy-ing	nitrify-ing	sulfur-oxid-izing
						peptone	ammoni-um hypo-phosphate	aero-bic	anaer-obic			
1	Silty-sandy	0	+	+	+	+	+	+	−	+	−	+
2	Clayey	1.0	+	+	+	+	+	+	−	−	−	+
3	The same	2.5	+	+	−	+	+	−	−	−	−	+
4	" "	5.0	+	+	−	+	−	+	−	−	−	+
5	" "	10.0	+	+	−	−	−	−	−	−	−	+
6	" "	12.0	+	+	+	+	−	−	−	−	−	+
7	" "	22.5	+	+	−	−	−	−	−	−	−	−

A similar decrease in number of bacteria with depth in a sequence of marine sediments has been noted also in the works of Butkevich (1932), Kolesnik (1938), Dianova and Voroshilova (1941), Kriss (1959), and other investigators.

This phenomenon is primarily associated with the compaction of sediments and the decrease of free moisture in them. According to Veber (1956a, 1956b), moisture is everywhere high (up to 70-80%) in organogenic muds. However, within the interval 0-2.0 m, the moisture in the Caspian and Azov-Black Sea muds falls to 18-48%. There is also a certain significance in the change in quantity and composition of organic material. According to Veber and Shabarova (1953), during diagenesis in organogenic muds the amount of easily hydrolyzed organic material (albumins and carbohydrates) decreases, and there is a relative increase in lipoids and so-called nonhydrolyzing residue, consisting of substances not easily available for assimilation by microorganisms; the total amount of organic material decreases.

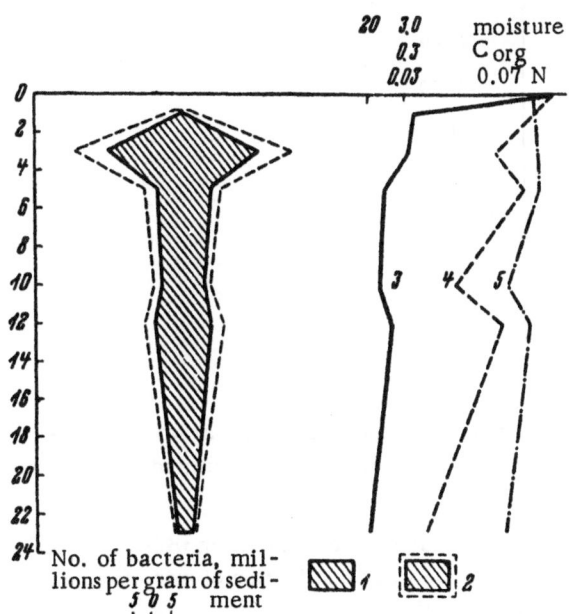

Fig. 1. Distribution of bacteria, moisture, C_{org}, and N in the section of Caspian sediments. 1) Number of bacteria per gram of moist sediment; 2) number of bacteria per gram of dry sediment; 3) moisture; 4) organic carbon; 5) total nitrogen. Ordinate axis represents depth in meters.

An investigation of the deeper horizons of the ancient Caspian (Khazary stage— Middle Quaternary) in the interval 165.3-583.8 m, made in our laboratory, has demonstrated that there is a great scarcity of microbial life in this stage of diagenesis of the sediments.

The maximum number of bacteria per gram of rock does not exceed 3100; in four samples out of nine bacteria could not be found, either by inoculation on special media or by direct microscopic counting (Table 2). The number of bacteria in the remaining five samples proved to be very uniform. Saprophytic bacteria (facultative anaerobes) were found in these samples; in two samples, weakly active denitrifying bacteria were present, and in only one sample, from a depth of 583.9, were glucose-fermenting and sulfate-reducing bacteria found.

A comparative analysis of the upper and lower parts of the section of Caspian sediments is very instructive. Quantitatively and qualitatively the microflora differ sharply at the various stages of diagenesis of the sediments. This is in keeping with the change in composition of the organic material in the sediments, the content of easily hydrolyzed components decreasing with depth.

According to Gorskaya (1956), the quantity of these easily hydrolyzed substances in the organic mass of the present-day sediments of the Caspian amounts of 46 to 77% on the average; in the Khvaly sediments these are notably less abundant (15-20%), and in the Khazary and Baku sediments they are practically absent. At these later stages of diagenesis there is generally little humic material; insoluble compounds predominate (in the organic mass), and in a number of places there is a marked increase (up to 20-30%) in the content of bituminous components.

The nature of the microflora in the Khazary sediments of the ancient Caspian is very similar to that described by Mekhtieva and Malkova (1958) for the Tertiary rocks of northern Ciscaucasia, which are also chiefly clay-silt and clayey deposits with a moisture content of 10 to 20%. These authors studied 240 samples of core from Eocene, Oligocene, Miocene, and Pliocene beds collected from depths ranging from 50 to 1000 m.

Live microflora were found in 85% of the samples; the forms were chiefly saprophytic, denitrifying, fat-splitting, and heptane-oxidizing bacteria. Less abundant forms encountered included ammonifying, glucose-fermenting, sulfur-oxidizing, and paraffin-destroying bacteria. Methane-, nitrifying and cellulose-destroying forms were absent entirely. The number of saprophytes in the rocks was on the order of 10^2-10^3 per gram of absolutely dry rock. A direct count under the microscope indicated from 10^4 to 10^6 bacterial cells per gram of rock.

The results of computing bacterial cells under the microscope and a determination of the number of heterotrophs by growths on culture media attest that the total number of bacteria in the rocks is much smaller than in modern marine sediments or in soils. The presence of bacteria of different physiological groups indicates that

TABLE 2. Number of Bacteria in the Section of Khazary Sediments of the Ancient Caspian (per gram of moist sediment)

Depth of sample, m	Type of sediment	Total no. of cells	Bacilli	Cocci
165.3	Clayey silt	0	0	0
186.5	Silty clay	0	0	0
260.8	Clay	1500	1500	0
316.5	Silty clay	3100	3100	0
350.0	Clayey silt	3050	1550	1500
440.6	Silty clay	1500	0	1500
534.5	The same	0	0	0
539.8	The same (from bacteria)	0	0	0
583.8	The same	3000	0	3000

various processes of transforming organic and mineral components of rock by microorganisms are at work in sedimentary rocks even at the present time. However, the scale of such processes is generally very limited, as attested to by the insignificant concentration of bacterial cells.

The results of microbiological investigations have been subjected to analysis and comparison with geological, geochemical, and chemical data, to seek light on any possible relationship between distribution of microflora and depth of rock, lithic type of rock, age of rock, and individual physicochemical factors.

A microbiological study of the section of a drill hole has shown that greater distance of an investigated horizon from the earth's surface has no effect on the number or kind of microflora (except for rocks taken from the zone of intense water exchange). It has not been possible to establish any relationship between microfloral content and stratigraphy of the rocks.

Nor has any direct relationship between content of organic carbon and nitrogen and the growth of microflora been ascertained, although in some parts of the section it has been possible to recognize a relationship between the number of bacteria and nitrogen, especially at low concentrations of nitrogen. A clear interconnection between natural moisture and number of bacteria in the rock has been noted.

When we compare the natural moisture of a rock with data from a quantitative computation of microflora on a culture medium (Fig. 2), we may then observe a very distinct relationship between these factors. An apparent lack of agreement in individual samples may be explained by differences in grain size and mineral content of the rock, and consequently, by differences in hygroscopic character. Apparently, because of insignificant amounts of free water in the rocks, the principal factor determining and limiting the growth of bacteria is water.

In order to test this conclusion we conducted a small experiment to study the growth of bacteria in moistened rocks. Ground rocks (clay and silt) were placed in sterile jars with ground-glass covers, moisture was added to 50% by weight of the rock (doubly distilled sterile water), and the tested material was kept at room temperature. Daily computations of the number of saprophytes were made by examining the proper cultures on culture media. At the beginning of the experiment the number of bacteria was on the order of 10^2-10^3, but at the end of a day it has increased to 10^5, and at the end of two days to 10^6. The experiment showed that when moisture was present the bacteria grew actively by using the nutrient substances found in the rock, and that the development of bacteria in the rock was limited merely by inadequate moisture.

The microflora in the deep horizons of the ancient Caspian is very similar to that in clay-silt and clayey varieties of Tertiary formations in northern Ciscaucasia, a fact apparently due primarily to a like content of free moisture in the ancient sediments and in the Terriary rocks (10-20%).

The oxidation-reduction status of the investigated rocks and sediments is in complete agreement with the character of the microflora found. In the present-day and ancient Caspian sediments, where a reducing environment is dominant, strictly anaerobic microorganisms are present side by side with facultative anaerobes, the former requiring low oxidation-reduction values for their development. In the Tertiary rocks, which are characterized by positive values of oxidation-reduction potential, the microflora are chiefly facultative anaerobic forms, thriving through a wide range of oxidation-reduction values and at various partial pressures of oxygen; aerobes are also present.

Conclusions

1. The development of microflora in clayey and silty marine sediments of the present day and of earlier times is determined primarily by the presence of free moisture in the sediments. The kind and quantity of organic material and of nitrogenous compounds, and also the oxidation-reduction potential are subordinate factors.

2. In the process of diagenetic conversion of sediment to rock, in connection with the great loss of moisture and the decrease in amount of easily hydrolyzed organic material, there occurs a marked impoverishment of microflora in the sediment, both in kind and in numbers. In regard to this, the role of microorganisms in transforming organic and mineral components of the sediments declines.

3. The character of the microflora in present-day and ancient clayey and silty marine sediments corresponds to the oxidation-reduction potential of the substrata.

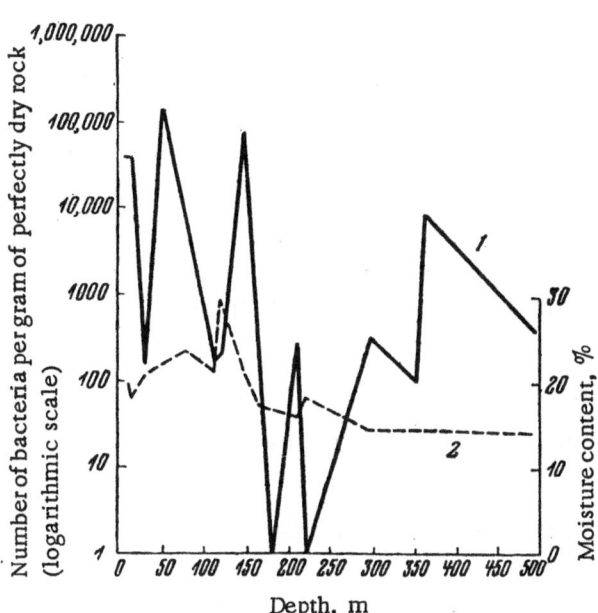

Fig. 2. Number of bacteria and the moisture of rocks in the section of drill hole 7 (Tashla). 1) Number of saprophytic bacteria; 2) moisture content of rocks, %.

LITERATURE CITED

Beerstecher, E. 1954. Petroleum Microbiology, Van Nostrand, Princeton.

Butkevich, V. 1932. "Technique of bacteriological investigation and some data on the distribution of bacteria in the water and bottom sediments of the Barents Sea." Trudy Gos. okeanograf. inst., 2, No. 5.

Dianova, E. and Voroshilova, A. 1941. "Bacterial profile of marine and lacustrine sediments as an indication of erosion and age." Doklady Akad. Nauk, SSSR, 30, No. 3.

Gorskaya, A. I. 1956. "A study of the organic material in recent marine sediments." in the Collection: Accumulation and Transformation of Organic Material in Recent Marine Sediments [in Russian], Gostoptekhizdat.

Kolesnik, S. L. 1938. "Microbiological investigation on the sediments in the southern and middle parts of the Caspian Sea." Komis. po Kompl. izuch. Kaspiiskogo morya, Materials on the Hydrobiology and Lithology of the Caspian Sea [in Russian], No. 5, Izd. AN SSSR.

Kriss, A. 1959. Marine Microbiology [in Russian], Izd. AN SSSR.

Mekhtieva, V. 1956. "A study of microflora in the present-day and ancient Caspian." in the Collection: Accumulation and Transformation of Organic Material in Recent Marine Sediments [in Russian] Gostoptekhizdat.

Mekhtieva, V. L. and Malkova, S. N. 1958. "Material on the microbiological characteristics of Tertiary and Quaternary deposits in northern Ciscaucasia." Trudy VNIGNI, No. 11.

Rittenberg, S. 1940. "Bacteriological analysis of some long cores of marine sediments." J. Marine. Res., 3, p. 191.

Veber, V. V. 1956a. "Accumulation of organic material in sediments." in the Collection: Accumulation and Transformation of Organic Material in Recent Marine Sediments [in Russian], Gostoptekhizdat.

Veber, V. V. 1956b. "The transformation of organic material." in the Collection: Accumulation and Transformation of Organic Material in Recent Marine Sediments [in Russian], Gostoptekhizdat.

Veber, V. V. and Shabarova, N. T. 1953. "The change in organic material of recent marine sediments during the transformation of the material." Trudy Mosk. filiala VNIGRI, No. 3.

ZoBell, C. E. 1946. Marine Microbiology, Baltimore.

ZoBell, C. and Anderson, D. 1936. "Vertical distribution of bacteria in marine sediments." Bull. Amer. Assoc. Petrol. Geologists, 20, p. 258.

THE LOWER BOUNDARY OF THE BIOSPHERE IN EUROPEAN USSR ACCORDING TO REGIONAL GEOTHERMAL INVESTIGATIONS

V. A. Pokrovskii

(Laboratory of Hydrogeological Problems, Academy of Sciences, USSR, Moscow)

In the concept of the biosphere, V. I. Vernadskii (1926, 1937) included the troposphere, the hydrosphere, and the upper part of the shell of sedimentary rocks, within which living substances are very significant in the geochemical processes of migration of chemical elements.

Except for rare exceptions, the mineralization, chemical composition, and pH of ground water cannot present any serious obstacle to the growth of microorganisms.

Microorganisms are most sensitive to temperature conditions. A study of the literature on the effect of temperature on the activity of microorganisms shows that the critical temperature for growth under normal conditions should be somewhere between 70 and 80°. However, in studying individual examples of microorganisms living in water at temperatures near 100°, and in view of the experimental investigations of ZoBell (1958), we have the critical temperature for the development of life in the deep zones of the earth's crust at 100°; and we may, with adequate grounds, consider the depth of the 100° isotherm to be the lower boundary of the biosphere.

In 1958-59, at the Laboratory of Hydrogeological Problems of the Academy of Sciences, USSR, the author and his co-workers B. G. Polyakii and V. I. Naidenova generalized the geothermal data of European USSR obtained from drilling and geophysical investigations during prospecting-exploratory work for oil, gas, coal, and other mineral deposits. Some of the geothermal data were obtained by special subsidiary drilling. In all, about 4500 thermograms were collected from European USSR, and more than 700 individual measurements of temperature were made with electrical and mercury maximum-reading thermometers, for more than 5000 deep holes, giving values of temperature conditions to depths reaching 3500 m. These data were used to construct a series of geothermal maps on a scale of 1 : 4,000,000 for European USSR, consisting of maps and cross sections for depths of 500, 1000, and 1500 m below the surface, isothermal contours on the crystalline basement, and maps showing depth to the 50° isotherm. In addition, geologic-geothermal profiles were made through all of European USSR along nearly east-west and north-south lines.

An analysis of all these geothermal data and of the prepared maps and profiles gives a rather complete picture of the temperature regime in the sedimentary cover and on the basement, and it permits one to divide European USSR into geothermal regions; in so doing we obtain three geothermal provinces:* a province of low temperatures, within which the temperature at 1000 m below the surface does not exceed 20°; a province of moderate temperatures, ranging from 20 to 30°; and a province of high temperatures, 30° and above.

---

* By geothermal provinces we here mean regions with relatively uniform geothermal conditions. The criterion for distinguishing a geothermal province is the temperature interval characteristic of a depth of 1000 m, at which almost no surface factors (climate, relief, water mass, cold water percolating down from the surface, dynamics of ground water, etc.) are capable of noticeably disturbing the natural thermal field of the earth.

Each of the designated geothermal provinces is generally characterized by characteristic geologic structure. The province of low temperatures occupies the region of the crystalline shields (Ukrainian and Baltic) and underground extensions of the basement (Kursk-Voronezh, Byelorussian-Lithuanian). The province of moderate temperatures corresponds to the platform zone with Precambrian crystalline basement and sedimentary cover of Paleozoic and Mesozoic rocks. The province of highest temperatures includes the areas of young Alpine depressions and basins with deep (to 8-9 km) Hercyanian folded basement and thick sedimentary cover of Mesozoic sandy-clay sediments. The boundaries of the geothermal provinces are shown on the map (see figure).

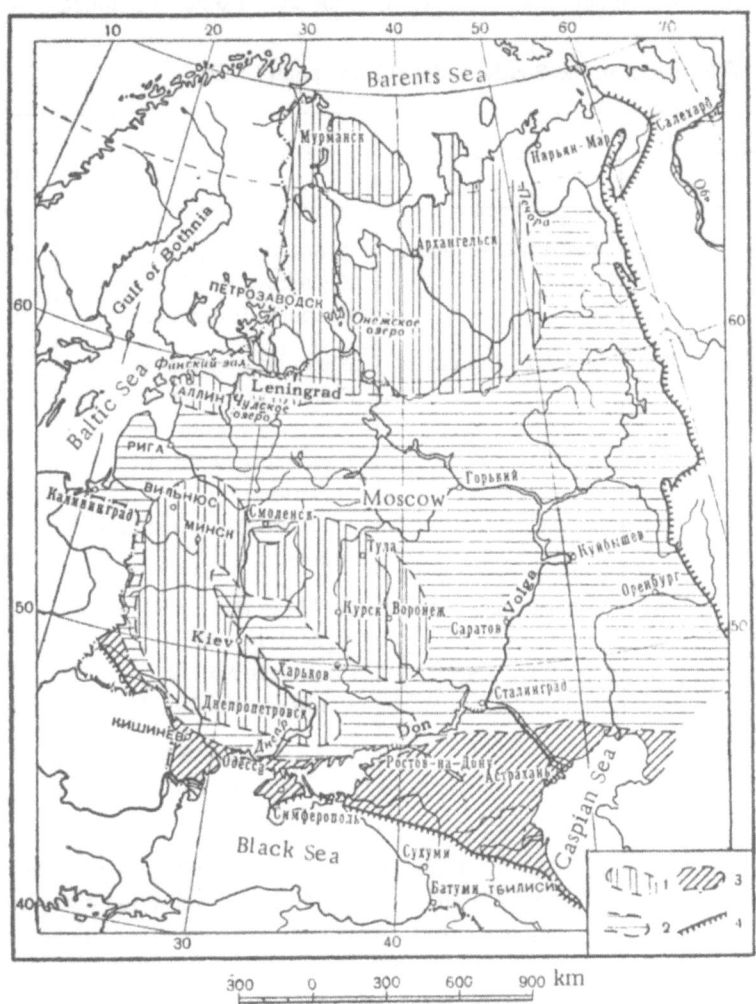

Map showing geothermal subdivisions of European USSR (prepared by
V. A. Pokrovskii, 1958). 1) Geothermal province of low temperatures;
2) geothermal province of moderate temperatures; 3) geothermal
province of high temperatures; 4) boundaries of mountainous structures.

In keeping with the temperature range dominant in each of the designated geothermal provinces, the $100°$ isotherm is found at a definite depth interval. Within the province of low temperatures, including the Kola Peninsula, the Ukrainian crystalline shield, and the Byelorussian-Lithuanian and Kursk-Voronezh salients (see figure), the $100°$ isotherm, by extrapolation, should lie at depths 10 to 15 km below the earth's surface. Within the province of moderate temperatures, corresponding to the extensive region of the Russian platform, the $100°$ isotherm occurs at depths ranging from 2.9 to 5.5 km. In the Dnepr-Donets basin, where the geothermal level is somewhat higher, actual measurements (Kupyansk exploratory hole 1) show the $100°$ isotherm to lie at a depth of 2.9 km. In the Caspian basin the $100°$ isotherm, by extrapolation, should lie at depths of 5 to 5.5 km, in the Volga-Ural region at depths from 3.2 to 4 km. In the remaining part of the Russian Platform the $100°$ isotherm lies at depths ranging from 4 to 5.5 km.

Within the geothermal province of high temperatures, including the regions of the Ciscaucasian, fore-Dobrogean, and Ciscarpathian depressions and of the Black Sea basin, which form the southern and southwestern frame of the Russian Platform, the 100° isotherm, in keeping with the dominant high temperatures, is found at comparatively shallow depths and is well delineated by temperature measurements in drill holes (see table).

From the data in the table it follows that the 100° isotherm lies at depths from 1.5 to 2.9 km in the regions of young Mesozoic depressions. But in small individual districts, representing discharge centers of deep thermal waters, the 100° isotherm is found near the earth's surface. Thus, at the Bragunskii thermal springs (Terek Range, eastern Ciscaucasia), which have a temperature of nearly 90°, the 100° isotherm is found at a depth of 300-400 m below the surface.

Depth to 100° Isotherm in the Southern Part of European USSR

Structure	Area	Drill hole	Depth to 100° isotherm
Ciscausian depression	Western Ciscaucasia	Tverskaya No. 1	2550
		Klyuchevskaya No. 16	2240
		Novo-Dmitrovskaya	2130
	Central Ciscucasia	Aleksandrovskaya No. 14	1680
		Chkalovskaya No. 1	1550
		Adu-yurt No. 35	2414
		Aleksandreya No. 1 (exploratory)	2750
		Artesian No. 1 (exploratory)	1960
	Eastern Ciscaucasia	Bozhigan No. 1	2900
		Mozdok No. 10	2760
		Ozek-suat No. 12	1805
		Vladislavovka No. 19	1880
Black Sea Basin Ciscarpathia		Novoslovskaya No. 7	1550
		Danilovskaya No. 1 (exploratory)	2300

Beginning with the data on depth of the 100° isotherm in European USSR, it is possible to indicate rather precisely the position of the lower boundary of the biosphere. As follows from the cited data, the depth of the lower boundary of the biosphere changes rather systematically, depending on the nature of the geologic structure and on the geothermal regime. Within the regions of greatest chilling—the Baltic and Ukrainian shields, the 100° isotherm and, consequently, the lower boundary of the biosphere, should descend to depths ranging from 10 to 15 km. On the platform, the boundary of the biosphere, although it may approach the surface of the earth, still is found at considerable depths (3.2 to 5.5 km). And lastly, in the regions of the young Alpine depressions along the southern and southwestern border of the Russian Platform, the lower boundary of the biosphere rises sharply and is found at depths ranging from 1.5 to 2.9 km, and in individual localities, at discharge centers of high-temperature waters, it approaches the surface of the earth still closer (300-400m).

Regional geothermal investigations may be very significant for microbiological studies, since they permit rather accurate prediction of temperature conditions at various depths and they define the position of the lower boundary of the biosphere, something not possible by other methods.

The author wishes to express his thanks to L. D. Shturm for very valuable advice during preparation of the present paper.

Conclusions

1. Data treated include 4500 thermograms and 700 individual measurements from more than 5000 deep drill holes in European USSR.

2. The data obtained make it possible to make geothermal subdivisions into districts and to delineate three geothermal provinces.

3. The province of low temperatures includes the Kola Peninsula, the Ukrainian crystalline shield, and the Byelorussian-Lithuanian and Kursk-Voronezh salients. The 100° isotherm lies at depths ranging from 10 to 15 km.

4. The province of moderate temperatures includes the Russian Platform, with the 100° isotherm lying at depths ranging from 2.9 to 5.5 km.

5. The province of high temperatures includes the Ciscaucasian and Ciscarpathian depressions and the Black Sea basin, where the 100° isotherm lies at depths ranging from 1.5 to 2.5 km, but in some localities reaches the earth's surface.

LITERATURE CITED

Vernadskii, V. I. 1926. The Biozone [in Russian], Leningrad.

Vernadskii, V. I. 1937. "Limits of the biosphere." Izv. AN SSSR, seriya geol., No. 1.

ZoBell, C. 1958. Ecology of Sulfate Reducing Bacteria. Producers Monthly, 1958, v. 22, N 7, pp. 12-29.

THE EFFECT OF MICROFLORA IN THE THIRD BED
OF THE YAREGA DEPOSIT ON CHANGES IN COMPOSITION
AND PROPERTIES OF THE OIL

I. L. Andreevskii

Oil Mine (Nefteshakhta) No. 1, Ukhta

The Yarega oil field is situated in the Ukhta region of the Komi ASSR and exploits the third oil-bearing bed of Middle Devonian strata, lying at a depth of 120-210 m.

Structurally this field occurs in the gently inclined Ukhta doubly plunging anticline, on which the limbs dip at angles ranging from 3 to 40°.

High-viscosity oil, complete absence of the benzene fraction, and low gas pressure in the bed make it practically impossible to work this bed from the surface. Therefore, since 1939 the bed has been worked by the shaft method, by means of three shafts, at first by underground boreholes drilled to a depth of 30-40 m from chambers in the overlying horizons, and then by inclined and horizontal boreholes from worked-out chambers in the bed itself (inclined system). Systematic observation on the properties and composition of the oil from 1940 to 1958 has shown, on the basis of nearly 1500 analyses, that the viscosity and flash point have diminished regularly, the first from $42°E_{50}$ to $29°E_{50}$ as a minimum, and the second to a considerably less degree, i.e., from 120 to 114° as a minimum (Andreevskii, 1959). The change to the inclined system of working the bed in 1951 to some increase in the values of these constants because of changes in the method of drainage; after this the values were again observed to drop. Observations of individual inclines also showed a decrease in viscosity from $13°E_{75}$ to $8.8°E_{75}$ during the interval from 1952 to 1955.

Similar results were obtained during observations over a course of 10 years on changes in the properties of oil kept in sealed iron vessels, and also at elevated temperatures.

Such changes in the properties of oil, lying outside the limits of error of any determinations, and observed under conditions that exclude the introduction of lighter oils into the bed, might be explained by the catalytic action of vanadium in the oil, by the effect of radioactive salts, or by the action of microflora in the third bed.

The absence of any changes in the properties of the oil at temperatures above 100-110° leads us to believe that the most likely cause of the changes is activity of petroleum microflora aroused at the time of exploitation of the bed.

Samples of oil, water, and oil-bearing sand, taken directly from the bed, under almost completely sterile conditions (after the rocks break off at the bottom of the hole), mostly have contained sulfate-reducing and other, indeterminate, species of microbes, which grow on Van Delden, Giltay, and Tauson media. The growths on Van Delden medium give off a sharp unpleasant odor, more like mercaptan than hydrogen sulfide.

It should be noted that the samples of oil sands, as well as the samples of water and oil that flowed out of fractures in the bed at the oil "mine," were collected under almost completely sterile conditions, since they were made immediately after the rock was broken open, i.e., under conditions eliminating all possibility of preliminary contamination of the rock by the drilling instrument, drilling mud, pipes, and so forth.

A more detailed study of the changes in composition of the oil after it has been held above the formational water has shown that, despite a relative constancy of composition of all the oil, its separate fractions undergo extensive changes. The quantity of naphthenes increases for the 250-400° fraction 11.5, 22.3, and 35.3%, and the amount of methane hydrocarbons decreases 7.8, 18.1, and 25.4%. A change in specific gravity on the order of 0.01 was observed for almost all fractions.

Experiments for observing the change in properties of the oil when stored above complete and incomplete nutrient media (i.e., under conditions of "nitrogen starvation"), have shown that the asphalt-tar constituents in the oil held over media without nitrogen drop to 68-69% as compared to 75% for oil held over the complete nutrient medium. The sulfur content also decreases from 1.17 to 0.8% and the specific gravity drops from 0.9440 to 0.9400.

In comparing the composition of the mineral content of the formational water with the prescribed nutrient media, one readily discovers that the first chiefly lacks nitrates, sulfates, and phosphates, and, consequently, the conditions of growth of the microflora in the bed may be considerably improved by artificially introducing these salts into the bed. It may also be assumed that by consciously excluding nitrates from the nutrient substances we create conditions that stimulate utilization, by the microflora, of the heavy, tarry, nitrogen-bearing components of the oil. The destruction of the tarry components of the oil, even if but partial, and the separation of a certain quantity of free nitrogen because of the meager use of the substances by microflora cannot but improve the yield of oil from the bed.

A comparison of the data we have obtained with the results of work by ZoBell (1946), Ékzertsev (1958), and others furnishes grounds for stating that the change in the composition of oil in an anaerobic environment occurs by means of bacterial activity.

The biochemical processes leading to change in composition and properties of oil must be a potent factor in increasing the yield of petroleum by beds through decrease in quantity of tarry components in the oil and increase in formational pressure by separation of free gases as products of the life activity of microflora.

Further study of the chemistry of these processes permits one to isolate more effective species of microorganisms, or what is more probable, to obtain mixed cultures of microorganisms and to discover the most favorable conditions for their existence in order to guarantee the maximum desired change in the oil.

The first attempt to use microorganisms commercially for increasing the yield of oil from a bed, as is known, was made by ZoBell. He believes that one of the basic functions of bacteria is desorption (setting free) of oil in a bed, a function that is effected by several simultaneously acting processes. One of these is the solution of carbonates by carbonic and organic acids generated by bacteria, with oil being set free and draining away along expanded channels.

In the American variant of bacterial treatment of oil beds by the introduction of specialized microflora (with artificially inoculated properties), there has been inadequate consideration of the variability and adaptability of the microorganisms to their new conditions of existence.

It is very possible that cultures in the laboratory will beautifully decompose tarry components of oil and will decrease the viscosity and the specific gravity, as pointed out by ZoBell, whereas in a formation, which differs considerably from conditions in the laboratory, the organisms may not be able to adapt themselves.

Apart from the American variant, the bacterial method of treating beds may be approached better by using microflora already living in the beds.

The microflora in a bed may be activated either by introducing a complete nutrient medium into the bed or to install a "nitrogen starvation" regime, i.e., introduce media containing no nitrogenous components. There is basis for assuming that the more effective procedure will be that requiring the microorganisms to use nitrogen from the oil, chiefly from the heavy tarry components.

The temperature of the bed is also a factor capable of influencing biological processes to a considerable degree.

A constant temperature in the reservoir rock is favorable for the growth of microflora, but the average temperature of the third bed (8-10°) is somewhat below the optimum temperature for such growth; this optimum temperature is generally between 20 and 30°.

The oil in the third bed, being heavy and tarry, is favorable for the development of microflora. The absence of the benzene fractions, which penetrate the cell walls of microorganisms, is also a positive factor.

The specific gravity of oil in the third bed is 0.9478, the viscosity $7-9°C_{75}$. The oil contains 13.5% aromatic hydrocarbons, 4.4% paraffins, and 82.1% naphthenes.

The content of sulfur in the oil is 0.98%, of nitrogen 0.42%, of paraffin 0.69%, and of asphaltic tars 56.8%.

The water in the third bed contains no admixture that has proved detrimental to the growth of microorganisms, if we neglect the high content of radioactive elements, the effect of which must be discussed separately.

For abundant development of microflora it is necessary to introduce separately into the formational water sulfates, phosphates, and nitrogen-bearing substances. The reaction of the medium is favorable (pH = 7.2-7.4), but when there is extensive evolution of acidic products through the activity of the microorganisms, it is necessary to add alkalies.

The gas in the third bed contains no impurities that might retard the growth of microflora; it consists chiefly of methane, which constitutes as much as 94-96% of the total.

A brief examination of the third bed of the Yarega deposit shows that the reservoir rock is a favorable environment both for stimulating the growth of native microflora and for introducing specialized species of microorganisms.

During industrial testing, the selection of the experimental section of the bed is of great significance in effecting bacterial treatment of the bed, since a uniform supply of nutrient medium on the surface of the sandstone will depend on the properties of the reservoir rock and on the method of supplying the medium.

The selected section should be, so far as possible, uniform in composition throughout the productive part of the bed in order to guarantee comparatively uniform oil saturation and permeability.

In this section it is necessary to have a sufficient number of drill holes, both for supplying the nutrient solution and for observing the results of the experiment.

Since the main objective of bacterial treatment, in contrast to flooding, is not to force the oil out by water, but to furnish the bed with a nutrient medium, the task is considerably lightened, and the expenditure of water is decreased.

The quality of the water used for preparation of the nutrient medium should be such that, so far as possible, the reaction of the environment is not changed and the proportions of the principal components in the formational water remain unaltered.

The concentration of the nutrient substances and their selection should be such that the pores of the bed will not be sealed up. From this point of view the formational waters of the "mine" are most suitable.

The amount of liquid supplied should correspond to the volume of the cavities formed in the bed after the oil, gas, and water are drained out.

In considering the average duration of bacterial treatment to be six months, a computation of the total volume of cavities to be filled reveal that it is necessary to pump 60-70 m^3 of solution per day, as a maximum.

The expenditure of nutrient substances will, on the whole, depend on the variant of bacterial treatment adopted.

Since the final selection may be made only after tests of the proposed variants at the "mine," below we present data on the computation of nutrient substances per cubic meter of solution per day and for the entire duration of the treatment, calculated for the maximum variant (see table).

These data show that the total expenditure of nutrient substances does not exceed 84-155 kg/m^3 per day or 12.6-23.3 tons for the entire treatment. Since almost all the nutrient substances are basic components of ordinary mineral fertilizer, one might expect that the cost will not exceed the cost of mineral fertilizers, i.e., approximately 2.0-2.5 rubles per ton.

The amount of oxygen, equal to 6.027 g per m^3, will come to 360-420 g per day for a daily delivery of 60-70 m^3 of water.

The amount of material needed to neutralize the acidic reaction products is determined by direct tests.

The treatment is carried out in the following way: water of the third bed, containing the added amounts of necessary nutrient substances, is pumped under a pressure that guarantees an introduction of 60-70 m^3/ day into the bed.

To cut down on loss of nutrient solution the drill holes should be covered and the oil that accumulates in the shafts should be periodically drawn off.

Expenditure of Salts during Pumping of Nutrient Medium into Bed (kg)

Salt	"Nitrogen-starved" medium			Ordinary medium		
	per m^3	per day	for 6 mos.	per m^3	per day	for 6 mos.
Ammonium sulfate	None	None	None	1.2	71.4	10,710
Calcium sulfate	0.6	42.0	6300	0.6	42.0	6300
Magnesium sulfate	0.3	21.0	3150	0.3	21.0	3150
Ammonium phosphate, monosubstituted	0.15	10.5	1575	0.15	10.5	1575
The same, di-, trisubstituted	0.15	10.5	1575	0.15	10.5	1575
Iron sulfate	0.012	0.84	126	0.012	0.84	126

This discussion refers only to the first preliminary stage of the whole complex operation of introducing the bacterial method of treating petroleum beds and petroleum itself in the practice of oil production.

It is necessary to emphasize once more that the novelty and complexity of the problem do not permit us to furnish a complete account of bacterial treatment of beds; much tedious work is still necessary, not only to discover the systematic pattern of the bacterial processes and the means of directing them, but also to develop a number of techniques not yet firmly established. As an example, it is sufficient to recall that even the balance of materials for destroying oil by microbes has not yet been ascertained, and, because of this, the error in disturbing the balance is as great as 25-30%.

Contradictory data from the literature are explained not only by errors in the experiments, but also by lack of appreciation of the extraordinary variability and adaptability of bacterial processes and by failure to realize the dependence of these processes on changes in the surrounding medium. In particular, such contradictory statements may be found in the observations of various authors, some maintaining that the oil becomes heavier, some lighter. Apparently both phenomena are but different sides of the same process, which takes place under various conditions not yet completely studied.

If, during the first "accumulative period" in the development of petroleum microbiology, each observation, each "recording of facts" has enriched the science and moved it forward, it is now necessary, at the present stage, to use bacterial processes for active creative work in nature.

Only by removing a process from its natural setting, by underestimating the connection between bacterial processes and their environment, and by failure to consider the variability and adaptability of these processes can such talented scientists and experimentors as ZoBell have failed to succeed in their attempts to use bacterial processes on an industrial scale to increase the yield of oil and to improve the quality of the oil.

We hope that the joint work of Soviet petroleum experts, microbiologists, biochemists, and geologists lead to the development of such means of bacterial treatment of oil and oil beds as to again point up the possibility of using bacteriology not in the interest of war, but in the interest of peace.

Conslusions

1. Laboratory experiments have shown that, in storing oil above formational water or over Tauson nutrient medium without organic material, in an anaerobic environment, naphthenic hydrocarbons increase and methane hydrocarbons decrease. The quantity of asphalt tars diminishes and the specific gravity of the Ukhta oil is lowered.

2. It is proposed that the yield of petroleum be increased from the bed by setting up experiments on activation of the microflora in the bed by pumping in a solution of sulfates and phosphates.

LITERATURE CITED

Andreevskii, I. L. 1959. "Means of using petroleum microbiology in the oil-producing industry." Trudy VNIGRI, No. 131.

Ékzertsev, V. A. 1958. "Study of the destruction of oil by microorganisms in anaerobic environments." Mikrobiologiya, 27, No. 6.

ZoBell, C. E. 1946. "Action of microorganisms on hydrocarbons." Bact. Review, 10, No. 1-2, p. 1-41.

BACTERIAL CHANGES OF OILS AND ITS COMPONENTS
IN ANAEROBIC ENVIRONMENTS

T. L. Simakova, Z. A. Kolesnik, N. V. Strigaleva, I. K. Voronova,
N. I. Shmonova, Z. S. Gerasyuto, and L. G. Andreeva

(All-Union Scientific-Research Geological-Prospecting Institute, Leningrad)

As the results of our previous investigations have shown (Simakova and others, 1958), the trend of change in oils depends not only on the chemical composition of the oils, but also on the kind and number of bacteria in the biocoenosis. To refine this position the present investigations follow from our belief in the necessity of studying the effect of the activity of biocoenoses, selected from the formational waters of various oil fields, on transforming some of the components of paraffin oils (Makhachkala and Tashkala).

Experiments for each individual biocoenosis were conducted in the following manner: 1.6 liters of formational water and 16 liters of sterile mineral medium were placed in 20-liter sterile bottles (the mineralization of the medium corresponded to the mineralization of the formational water, from which samples were collected for microbiological investigations). The material in the bottles was inoculated with an amount of 10% of the volume of the medium. The mineral medium was inoculated by accumulated cultures on mineral media containing oil or one of its components as the sole source of carbon. A six-day culture of sulfate-reducing bacteria was introduced with a modified Tauson medium containing the minimum quantity of sodium lactate (0.05%). As demonstrated by earlier experiments, the amount of sodium lactate used was that required by sulfate-reducing bacteria for a period of 5-6 days. The medium thus contained no organic material besides the oil components to be tested. The medium and the inoculating material, well shaken in the bottle, were introduced in the amount of 4 liters into a 7-liter round-bottom flask, after which the component of oil to be tested was added. The hydrocarbon fraction and the lubricating oils were used in quantities of 25-30 g for each 4 liters of medium; the hydrocarbons, tars, and asphaltenes were used in quantities of 2 to 5 g for each 2 liters of medium.

The remaining air in the flasks was removed by passing nitrogen through the flasks after preliminary cleaning of the nitrogen by an alkaline solution of pyrogallol.

During the experiment, studies were made in the aqueous medium on the dynamics of the quantity of bacteria, on the amount of residual carbon easily dissolved by organic material (the product of bacterial activity), on the concentration of hydrogen ions, and on the chemistry of the changes in the components subjected to bacterial activity. Physicochemical, optical, and luminescent methods were used.

The microflora for the experiments were obtained from formational waters at deposits in Tashkala, in the Staro-Grozny region, in Emba, and at Makhachkala.

In the work of 1958-59, the experiments were carried out with individually isolated components of oil: the hydrocarbon fraction, lubricating oils, tars, and asphaltenes.

In group composition, the lubricating oils in the experiments with biocoenoses from Emba and Tashkala had some tendency to become tarry. The tars, according to luminescent-component studies, changed by increase in molecular weight. The luminescent-component studies showed, in all experiments, a decrease in the content of aromatic hydrocarbons relative to the control. Optical studies revealed the appearance of oxygen-bearing compounds in the experiments with the Tashkala Staro-Grozny microflora. These were not detected in the control.

Hydrocarbons isolated in separate groups from the fractions that boiled off at 250-300° and at 540-500° from the Makhachkala oil and from the 350-400° fraction of the Tashkala oil changed dissimilarly.

Methane hydrocarbons, isolated by forming a complex with urea from the 250-300° fraction, suffered no change when acted on by the microflora from the Makhachkala deposit. According to optical studies, a decrease in the ratio of the CH_3 group to the CH_2 group relative to the control was noted in the methane hydrocarbons from the 450-500° fraction of the same oil and with the same biocoenosis of bacteria. Such a decrease in the number of CH_3 groups might attest to the dominant removal either of more branched hydrocarbons or of hydrocarbons of lower molecular weight.

As the results of optical studies have shown, the methane hydrocarbons separated by forming a complex with urea from the 350-400° fraction of the Tashkala oil changed by decrease in the ratio CH_3/CH_2 in experiments with all three biocenoeses of bacteria.

The most pronounced change was noted in the experiments with microflora from Emba.

The various results obtained during study of the influence of microorganisms on the change in mechane hydrocarbons separated with urea from the methane fractions were apparently due to differences in structure.

Methane hydrocarbons of normal structure were actually isolated from the 250-300° fraction; the methane hydrocarbons from the 350-400° fraction were more complex, the molecules containing naphthene structures.

Besides the above-indicated experiments, others were set up with chemically pure paraffin and with paraffin extracted from the 450-500° fraction of paraffin oil from the Makhachkala deposit. These experiments were carried out to shed light on the problem of whether paraffin may serve as the source of carbon for the examined microflora, particularly for denitrifying bacteria, which are widespread in the formational waters of various oil deposits. The results of the studies show that the bacterial process was most active, with the formation of large quantities of water-soluble substances, in experiments with a mineral medium containing nitrates.

Optical data has shown some qualitative change in the paraffin extracted from the 450-500° fraction in an experiment with a mineral medium containing no nitrates. A decrease in the group ratio CH_3/CH_2 was also noted.

In three other experiments no qualitative changes were detected in the paraffins themselves. Because of the similar values of the constant of the hydrocarbons in the paraffins, qualitative changes might not have been detected in the paraffins. It is possible that the entire range of hydrocarbons in these paraffins may have been destroyed, with water-soluble substances forming in their place. These latter were found in considerable quantities in the experiments.

Isomethane and naphthene hydrocarbons changed dissimilarly.

The changes in these hydrocarbons taken from the 250-300° and 450-500° fractions of the Makhachkala oil and acted on by a biocoenosis of bacteria taken from the same oil led to a decrease in methane and an increase in naphthene hydrocarbons.

The isomethane and naphthene hydrocarbons extracted from the 350-400° fraction of the Tashkala oil were altered by a decrease in quantity of naphthene hydrocarbons and an increase in quantity of methane hydrocarbons when acted on by three biocoenoses (Emba, Tashkala, and Staro-Grozny). Optical studies of the isomethane and naphthene hydrocarbons indicated a decrease in numbers of the CH_3 group in all experiments relative to the control.

Thus, during anaerobic oxidation bacteria may use both isomethane and naphthene hydrocarbons, and the alteration of these apparently depends on the structure of the hydrocarbons.

In all the experiments the changes in aromatic hydrocarbons, both from the 250-300° and 450-500° fractions of the Makhachkala oil as well as from the 350-400° fraction of the Tashkala oil, were characterized by a decrease in the quantity of bicyclic hydrocarbons.

These hydrocarbons from the 350-400° fraction of the Tashkala oil changed insignificantly in but one experiment— with microflora from Tashkala. The slight availabiliby to bacteria of the aromatic hydrocarbons extracted from the 350-400° fraction was apparently due not only to the structure, but also to differences in specific content of bacteria in the studied microflora.

Besides the experiments indicated above, others were carried out for study of asphalt-tar substances extracted from the Makhachkala oil through the activity of microflora from the same locality, and still others were made on separately extracted tars and asphaltenes from the Tashkala oil through the activity of the three biocoenoses of bacteria referred to.

In the first of these, the asphalt-tar substances were unavailable to the bacteria and no changes in elemental composition of the substances was detected. In the second, the elemental composition of the tars was altered by microflora from Emba, and the composition of the asphaltenes was altered by the biocoenosis from Tashkala.

In studying the dynamics of the quantity of bacteria of the investigated biocoenoses, it is necessary to note the quantitative relations of denitrifying to sulfate-reducing bacteria, since these determine generally the magnitude of the changes in the investigated components of petroleum. In experiments with the microflora from Emba, denitrifying bacteria were more abundant than sulfate-reducing bacteria, and the activity of these latter was greater than it was in the experiments with microflora from the Staro-Grozny region. Chemically the greatest change in the investigated components was observed with the biocoenosis of bacteria from Emba.

We repeatedly noted a favorable effect of some strains of denitrifying bacteria on the activity of sulfate-reducing bacteria (Kolesnik and Shmonova, 1957).

On the other hand, experiments of Lomova (Simakova and Lomova, 1958) have shown that the presence of large numbers of some strains of denitrifying bacteria depressed the activity of sulfate-reducing bacteria.

Studies on pure cultures have shown that the investigated biocoenoses differ among themselves only in specific content. These species are divided into varieties and strains of bacteria. Most of the cultures belong to the genus Pseudomonas and are facultative anaerobes. It is interesting to note their selective relationships to oils of different composition; this fact allows us to refer to different strains of isolated cultures.

We did not succeed in making strict, specific isolations of bacteria; but pure cultures that gain their necessary carbon not only from the easily soluble substances, but also from the hydrocarbon components of the oil are, it seems to us, specific groups adapting themselves to the special conditions of their environing habitat—formational water and oil.

The sulfate-reducing bacteria and one other culture, the species of which we could not identify, are anaerobic microflora.

Pure cultures of sulfate-reducing bacteria did not multiply in a mineral medium containing oil or its hydrocarbon components as the sole source of carbon. In combination with other groups of bacteria they take an active part in the alteration of petroleum. This phenomenon may be explained apparently by the oil becoming available to them only after other groups of bacteria have altered it somewhat. But it is possible that substances may be evolved in the biocoenosis of some groups of bacteria which activate sulfate-reducing bacteria, and these latter then begin to use the carbon in the oil.

Active denitrifying bacteria were noted buring the bacteriological studies of large quantities of formational water from the various oil deposits. And, naturally, the problem arose: From which compounds does a given group of bacteria acquire the necessary energy in an anaerobic environment, when nitrates are absent? Experiments by Kolesnik have shown that some strains of denitrifying bacteria may acquire energy in an anaerobic environment by fermentation of easily soluble oxygen-bearing substances—carbohydrates, polyatomic alcohols, and so forth.

Experiments have also shown that these pure cultures of denitrifying bacteria may not only exist but may multiply in the aqueous extract from oil in which neither nitrates or other mineral salts have been introduced. Apparently they use oxygen-bearing substances washed out of the oil, hydrocarbons that are soluble in water, and nitrogen from heterogeneous components.

In the investigations of the aqueous medium in experiments set up with the hydrocarbon fractions and with groups of hydrocarbons, volatile fatty acids were detected (formic and, possibly, isobutyric). In the experiments with tars and asphaltenes these fatty acids were not found. An accumulation of residual carbon appeared in all the experiments, the quantity depending on the degree of bacterial alteration of the investigated components of oil.

Conclusions

1. Biocoenoses of bacteria that grow in an anaerobic environment on mineral nutrient media with oil have been obtained. Formational waters from Tashkala, the Staro-Grozny region, Emba, and Makhachkala were used to isolate the biocoenoses.

2. Most of the bacteria in the biocoenoses belonged to the genus Pseudomonas, differing in specific content.

3. A study of the influence of the biogenic factor on the alteration of the hydrocarbon part and the tars and asphaltenes isolated from paraffin oils has shown that these substances change to some degree in an anaerobic environment through the biogenic factor. The scale of the alteration is small, but a definite trend in these processes has been noted.

4. The alteration of the components of oil depends not only on their chemical structure, but also on the kind and number of microflora present.

5. The interrelations among the physiological groups of bacteria in a biocoenosis are very important in the alteration of oil. The trend and activity of alteration of oil and its components depend on the number of bacteria and on the medium in which they live. In the experiments a relationship has been observed between the numerical ratio of denitrifying to sulfate-reducing bacteria and the degree of alteration of the investigated component of oil. Thus, in experiments with the biocoenosis from Emba, a great number of denitrifying bacteria activated sulfate-reducing bacteria, and the alteration of the investigated components of oil became more active than in experiments with the other biocoenoses.

LITERATURE CITED

Kolesnik, Z. A. and Shmonova, N. I. 1957. "A study of the changes in oil in an anaerobic environment due to the influence of bacteria from the genus Pseudomonas." Doklady Akad. Nauk SSSR, 115, No. 6.

Simakova, T. A. and Lomova, M. A. 1953. "A study of the microflora in the oil fields of Vtoroi Baku." Trudy VNIGRI, No. 117.

Simakova, T. L., Gorskaya, A. I., Kolesnik, Z. A., and others. 1958. "Nature of the change in oils in anaerobic environments due to the biogenic factor." Trudy VNIGRI, No. 128.

THE ROLE OF MICROORGANISMS IN PRODUCING
THE CHEMICAL COMPOSITION OF GROUND WATER

M. S. Gurevich

(All-Union Scientific-Research Geological-Prospecting Institute, Leningrad)

It has been less than thirty years since living organisms were discovered in the deep artesian waters of oil deposits. Since that discovery, data have been accumulating on the quantitative distribution, the morphology, and the physiology of the microflora living in the underground hydrosphere. Investigations of Ginzburg-Karagicheva, Isachenko, Kuznetsov, Shturm, and other microbiologists have shed light on many phases of microfloral life in ground waters. The works of Vernadskii, Kozlov, Sokolov, and Uspenskii have promoted clarification of the role of these organisms in the geochemistry of ground water, particularly under deposits of oil and gas. The significance of microflora as a factor in transforming the composition of natural water has also found expression in modern hydrochemistry (Alekin, 1935), and this factor is gradually coming to be appreciated in hydrogeological investigations.

There is now no doubt that the chemical composition of ground water may change not only by inorganic processes, but also through the activity of microorganisms. In keeping with this view, it is proper to distinguish biogenic and abiogenic processes of transforming the chemical composition of ground water. Both types of transformation characterize, in different degrees, the different stages of geochemical history of ground water in the biosphere, beginning with the stage of early diagenesis and ending with the final stage of late katagenesis of rock. A study of the role of microorganisms in producing the chemical composition of ground water is one of the tasks of hydrochemistry, and it embraces the geochemistry of biogenic processes taking place in ground water and the effect that these processes have on the migration of the chemical elements in the earth's crust. It is proper in this context to speak of the biochemical study of ground water, having in mind the objectives of biochemistry defined by Vernadskii (1935).

We shall try, using the presently known facts, to note the chief peculiarities of biogenic transformation of ground-water composition at the various stages of geologic history of its development. Our attention will here be focussed chiefly on the significance of biogenic reduction of sulfates, which is closely related to the biogenic conversion of carbon compounds, and on the significance of oxidation of sulfides with the formation of secondary sulfates.

Transformations of the Saline Composition of Ground Water by Biogenic Reduction of Sulfates

The formation of alkaline waters of sodium bicarbonate-chloride composition has been considered, by a number of authors, to be the consequence of reduction of sodium sulfates, with the parallel formation of sodium bicarbonate and the precipitation of calcium and magnesium carbonate, leading to the accumulation of soda in solution and to a high alkalinity of the solution.

Different treatments of this phenomenon (Sulin, 1948; Alekin, 1953; Stadnikov, 1957) are based on the biogenic nature of the reduction process of sulfates, originating either in the stage of muddy water or in ground water circulating through already lithified sediments. The most strongly alkaline waters are found in oil deposits, where, because of the abundance of organic material, sulfates are reduced most intensely.

According to Lindtrop (1947), the water composition at the water-oil divide, in wells at the Grozny oil field, is characterized over short periods of time by a gradual decrease in the SO_4^{2-} ion, with a simultaneous increase in the bicarbonate content (see Fig. 1).

Numerous observations by a number of authors attest to a decrease in the content of sulfates and an increase in the amount of bicarbonates as the oil deposit is approached. Data confirming this are available for the Apsheron Peninsula (Zhabrev and Khachkevich, 1951), the Volga-Ural region (Belyakova, 1956), and other places. As an example of this change in composition we may cite the data of Mitgarts relative to bed V at the Yuzhnyi Alamyshek deposit (Table 1).

TABLE 1. Change in Saline Composition of Water with Distance from Oil Deposit

Well No.	Distance from oil-bearing zone, kg	SO_4^{2-}, g/liter	HCO_3^{2-}, g/liter
141	−1.5	1.70	0.06
138	−1.1	0.54	0.02
139	At boundary	0.08	0.05
15	+0.25	0.001	0.07

*Outside boundary−; inside boundary of oil-bearing zone +.

The biogenic transformation of water composition by reduction of sulfates is intense in muddy waters, being, according to V. I. Vernadskii, one of the initial stages in the development of ground water. Investigations of Shishkina (1958, 1959a), made during experimental cruises of the ship "Vityaz'" in the Okhotsk and Bering Seas and in the northwestern part of the Pacific Ocean, showed that the reduction of sulfates in muddy waters rich in organic material was more intense than in muddy waters containing little organic material.

In a vertical section of muddy sediments the amount of sulfates in solution decreases with depth and the alkali reserve in the muddy water increases. At the same time a decrease in the content of calcium is observed, due to its precipitation as carbonate. In muddy waters rich in organic material, the quantity of SO_4^{2-}, in comparison with sea waters, decreases from 9.3 to 0.5%, but the alkali reserve increases from 0.39 to 7.36%. The general saline composition of the water in this latter case is sodium bicarbonate-chloride, practically free of sulfates (see Fig. 1). According to the same author (Shishkina, 1959b), in the muddy waters of the Black Sea the relationships among change in quantity of sulfates, size of alkali reserve, and content of organic material in the sediments hold only in general outlines, and there are a number of deviations. The sharpest of these peculiarities is observed in the muddy waters in the upper layer of the bottom sediments, about 40 cm thick. During the reduction of sulfates in the Black Sea muds, sulfate-free sodium bicarbonate-chloride waters are also formed. In the older bottom sediments (Novoevksinskie beds) sulfate-free waters of calcium-sodium chloride composition have been found.

How does one explain the difference in the features of biogenic transformation of the muddy waters in the Okhotsk, Bering, and Black Seas? What causes in the Black Sea have complicated the relationship between changes in sulfate content in the water of the alkali reserve and the over-all content of organic material? These questions can be answered, apparently, only by computing the activity of microorganisms, because, apart from the fundamental differences in the geologic history of these marine basins, differences in the biogeochemical conditions of sedimentation have been undoubtedly influential.

Turning to the composition of the muddy waters of the Caspian Sea, we are confronted with the already discussed facts concerning the change in the composition of the water during reduction of sulfates. According to Malyshek and Shaikhet (1959), muddy waters in the shallow part of the Caspian Sea, squeezed out of the bottom sediments, have a magnesium-sodium sulfate-chloride composition. The content of the SO_4 ion decreases with increase in depth of the muddy waters, whereas the content of HCO_3 increases; the amounts of calcium and magnesium also diminish. The fact that sodium bicarbonate-chloride waters were not detected in the investigated samples is explained by the authors by the shallow depths at which the muddy waters were collected, none being taken deeper than 60-70 cm below the surface of the sea floor.

In generalizing the discussed role of biogenic transformation in developing the composition of muddy sea water, one must conclude that bacterial reduction of sulfates changes the composition of sea water to sodium bicarbonate-chloride, which is typical of several oil deposits (see Fig. 1). During succeeding abiogenic transformation base exchange between sediments and water, the water acquires a calcium-sodium chloride composition.

Information on the composition of the muddy waters in the Black, Caspian, and Far Eastern Seas indicate that the connate water preserved in marine sedimentary rocks during early diagenesis is, depending on the sedimentary facies, at some stage of biogenic transformation.

In the upper layers of the muddy sediments of the Caspian Sea the connate water retains the composition of the sea water, being distinguished from the sea water merely by higher mineralization and a lower content of sulfates. In the muds of the Far Eastern seas, which contain considerably more organic material, the muddy waters acquire a sodium bicarbonate-chloride composition in the process of sulfate reduction.

The same process is actively taking place in the muddy waters of the Black Sea, but here it is considerably complicated by bacterial oxidation of hydrogen sulfide and by accompanying secondary biogenic generation of sulfates.

It is thus necessary to recognize that the studies of how the composition of connate marine water developed clearly attest to an important role of microorganisms in this process. Experimental investigations of Tageeva (1955) are also in agreement, attesting that bacterial anaerobic reduction of sulfates actually leads to the accumulation of bicarbonate ions in the solution and to the subsequent precipitation of magnesium and calcium carbonate in the sediment.

Independent of the studies on muddy waters, Stadnikov (1955), in a monograph devoted to argillaceous rocks, concluded that alkaline sodium bicarbonate-chloride waters in oil deposits were formed by bacterial reduction of the sulfates in sea water. In Stadnikov's view oil originates from thick accumulations of plant remains on the sea floor beneath the zone of hydrogen-sulfide fermentation. Below this zone, sea water is completely free of calcium; sulfates are reduced; and the carbonates that have formed precipitate in the sediment.

As a consequence of this circumstance, water and rock adjacent to oil at its site of original formation should contain no calcium that is capable of being replaced by sodium. Calcium salts appear in the water only during subsequent migration of the water, when the water comes into contact with rocks containing considerable quantities of calcium that may be replaced by sodium. During this migration, ground water is converted to calcium-sodium chloride water by a reversible type of exchange.

In assuming further that the oil and water migrate simultaneously, Stadnikov believes that alkaline waters, containing little or no calcium salts, occur near or at the site of the initial oil accumulation, and that calcium-sodium chloride waters are found at sites of secondary oil accumulations.

Without having in mind any critical examination of the given hypothesis, we should nevertheless note that it raises a number of objections concerning the formation of oil from the initial accumulation of organic material and concerning the subsequent migration of the oil as well as concerning the conditions under which the alkaline waters form. The assumption that sodium bicarbonate-chloride waters form only on the sea floor at places of muddy water and of sediments free of calcium salts, thus necessarily below the hydrogen-sulfide zone of sea water, is contrary to the facts. In the Black Sea, where the hydrogen-sulfide zone extends down to a depth of 2000 m, the muddy waters contain both calcium carbonate and calcium chloride

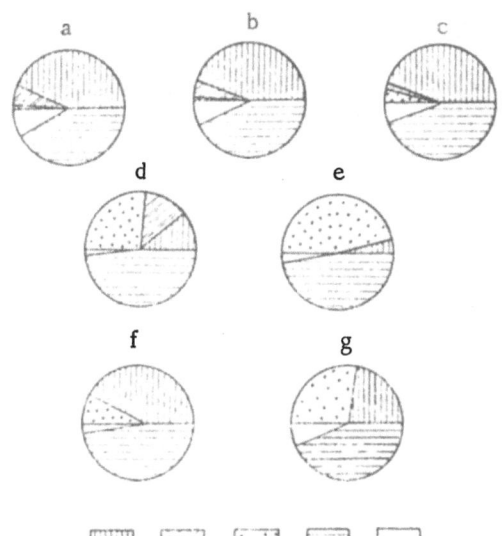

Fig. 1. Biogenic transformation of muddy and subsurface waters. a) Sea water. Muddy water (after O. V. Shishkina): b) mud poor in organic matter (Okhotsk Sea, depth of 1335 m); c) mud rich in organic material (Okhotsk Sea, depth of 964 m); d, e) water from oil deposits (after N. T. Lindtrop), Novogrozny region. Ground water from oil deposits on the Apsheron Peninsula (after V. A. Sulin): f) Il'icha Bay; g) Binagady. 1) Cl^-, 2) SO_4^{2-}, 3) alkali reserve of HCO_3 mud, 4) $Na^+ + K^+$, 5) $Ca^{2+} + Mg^{2+}$.

(Tageeva and Tikhomirova, 1957; Shishkina, 1959b). The muddy waters of lacustrine basins and land masses, judging from microbiological data (Kuznetsov, 1952), are characterized by the same peculiarities of biogenic transformation found in muddy waters of the sea. The transformation of connate waters by microorganisms in this circumstance is closely connected with the "microzonality" of the mud deposits (Perfil'ev, 1932).

From the example of oil fields, we might convince ourselves that the biogenic transformation of the composition of water takes place not only at the stage of early diagenesis, but also at the stage of late katagenesis, when water-bearing horizons have already developed and circulation of ground water has been established.

In artesian basins, sodium bicarbonate-chloride waters are found chiefly in the marginal zones of the artesian basins, bordering the fold-mountain structures, which are, at the same time, recharge zones of the artesian water. For example, such water is widespread in the southeastern part of the Western Siberian artesian basin, where it is confined to Tertiary and Mesozoic rocks.

The nature of this phenomenon is clear, since water of surface origin, advancing from the zone of recharge, carries sulfates with it, guaranteeing restoration of the store of this material, which is destroyed by reduction through oxidation of hydrocarbons in the dissolved organic material. With greater distance from the recharge zone and with greater depth in the artesian basin, the intensity of sulfate reduction and of the accompanying changes in saline and gaseous composition of the water weakens. A fundamental condition favoring biogenic transformation of underground water in the marginal zones of artesian basins, apart from the restoration of sulfate supply, is the distribution here of the most permeable sands and of polymict deposits in general. Polymict formations, associated in their formation with mechanically reworked erosional products of volcanic rocks from the zone of their erosion, give way in the deeper-water segments of the sedimentation basin to clayey deposits, through which water moves with much greater difficulty, thus complicating the penetration of sulfate waters from the zone of recharge.

It should be noted that, according to Strakhov (1951), the restriction of alkaline waters to polymict deposits is due to the presence of a number of reworked alkalic minerals in such deposits, the alkalies going into solution as the mineral is attacked, a process that undoubtedly favors the formation of alkaline water.

In speaking of the biogenic nature of formation of sodium bicarbonate-chloride alkaline waters, it is necessary to keep in mind that, for such water to form by biogenic means, not only is an abundant supply of sulfates necessary, but there must also be considerable organic material dissolved in the water. In waters of the oil and gas deposits of the Grozny-Dagestan oil region, a direct relationship has been observed between the content of organic carbon and the abundance of desulfurizing bacteria (Al'tovskii, Kuznetsova, and Shvets, 1958).

When there is an excess of organic material in the sediments and a deficiency of sulfates, and also when gypsiferous deposits are absent, the sulfates are reduced and practically eliminated entirely while organic material and gaseous hydrocarbons remain in solution. When there is a sharp dominance of sulfates (because of solution of gypsum by water moving from the recharge zone) and a limited supply of organic material in solution, the final stage of the same oxidation-reduction process produces oxidation of the organic material, and part of the sulfates is preserved. An example of this is the oil- and gas-bearing Tashkent artesian basin (second order), where Cretaceous rocks, according to Beler (1958), contain nitrogenous sodium sulfate-chloride bicarbonate waters.

Thus, the formation of sodium bicarbonate-chloride waters is controlled and determined by the well-known most favorable ratio of sulfates to organic material and by the very fact that sulfate-reducing microorganisms are present in the water. A change in the ratio of sulfates to organic material leads to the development of genetic varieties of this type of ground water, peculiar to different geochemical conditions.

The accumulation of sulfates in ground water is also promoted by periods of continental emergence, when sulfates move to great depths in waters of infiltration. It is obvious that this has taken place on the Apsheron Peninsula, at the boundary between the upper and lower units of the productive sequence. It is also necessary to keep in mind that sulfates may be abundant in muddy waters themselves that form in lacustrine and lagoonal basins.

There are ample grounds for speaking of definite stages in the biogenic reduction of sulfates throughout the geologic history of ground waters of the investigated composition (Gurevich, 1954, 1958a, 1958b). This circumstance cannot affect the observed stage relationship in the changes in composition of the water during its biogenic transformation. The dominant anions in this process change according to the following sequence: $Cl^- - SO_4^{2-} \rightarrow Cl^- - HCO_3^- - SO_4^{2-} \rightarrow Cl^- - HCO_3^-$.

In arriving at this conclusion one cannot ignore the probability that water of the same composition may form under other conditions, where abiogenic transformation is more important than biogenic. It is not altogether clear, for example, how important the role of microorganisms may be in the formation of sodium bicarbonate-chloride carbon-dioxide waters in the zones of Alpine tectogenesis, where free carbon-dioxide waters very likely penetrate to great depths. For such circumstances Tageeva (1958) assigns considerable importance to carbon dioxide as a factor promoting the transfer of calcium and magnesium bicarbonates from sand-clay rocks to ground water. However, keeping in mind the value of carbon dioxide as a solvent, one should also take into consideration the fact that at depths where the temperature does not exceed the critical value of $32°$ the carbon-oxygen compounds are chiefly the ions HCO_3^- and CO_3^{2-}. These depths, in keeping with the geothermal gradient, are 1-2 km below the earth's surface.

In this view of the marked effect of carbon dioxide on the carbonates in water-bearing rocks, Zhabrev and Khachkevich (1951) have determined the part played by the carbon dioxide that forms in waters of oil deposits by reduction of sulfates.

However, the data of Shishkina concerning the prime solutions in muds refute this position. During the reduction of sulfates because of the accumulation of carbon dioxide in solution and because of increased pH, the concentration of calcium and magnesium does not increase; it decreases rather as a result of precipitation of $CaCO_3$ in the sediment.

In considering the reduction of sulfates as a biogenic process, we start not merely from the fact that the process is closely associated with the activity of the appropriate bacteria. Experimental work of a number of investigators has shown that abiogenic, chemical reduction of sulfates is possible at high temperatures and that only the interaction between methane and sulfates, according to experiments of Grigor'ev (1954), leads to such reduction at temperatures of $100°$ and lower.

The foregoing facts and discussions compel us to conclude that the formation of alkaline sodium chloride-bicarbonate ground water, during biogenic reduction of sulfates by oxidation of organic material, is one of the most probable and widespread means of producing such water in the biosphere. The process is must extensive under the geochemical conditions securing a sufficient supply of sulfates, an excess of dissolved organic material, and an environment promoting the active life processes of microflora. From this point of view, water of the indicated composition, when specific gas and ion components are present, may be considered the "companion" (sputnik) of oil and gas deposits, genetically related to them in the conditions of origin.

Clearly we cannot exclude the possibility that water of similar composition may form by chemical means under fundamentally different geochemical conditions, as has been proposed in a number of hypotheses, the examination of which is a special problem.

Biogenic Sulfofication of Ground Water

In considering this question it is necessary to bear in mind that the attention of investigators evaluating the biogenic transformation of sulfur compounds in ground water has been focussed thus far on explaining sulfur-free water; the other side of the question, relating to the secondary biogenic accumulation of sulfates in ground water. has fallen from view. Meanwhile there are a number of physiological groups of bacteria, some of them autotrophic, that oxidize hydrogen sulfide to sulfur, and sulfur to sulfuric acid.

A study of the microflora in the ground waters of artesian basins in the Western Siberian lowland, Uzbekistan, Turkmenistan (Dutova, 1959) has shown that sulfur-oxidizing bacteria and denitrifying bacteria that oxidize sulfur are widespread in these deep artesian waters, even at depths exceeding 1.8 km. The presence of sulfur-oxidizing dinitrifying bacteria has been noted in the artesian waters of oil fields on the Russian Platform (Kolesnik, 1951), in the Grozny region, and in Dagestan (Al'tovskii, Kuznetsova, and Shvets, 1958).

The sulfate content and the acidity of ground water adjacent to ore deposits and coal are due to the oxidation of metallic sulfides contained in these deposits. A basic geochemical feature of the oxidation zone of sulfide deposits is the removal of sulfur from the ore body by oxidation and the solution of sulfides, leading to the formation of sulfates. The oxidation of sulfides and the formation of sulfuric acid in this situation is thought to be a purely chemical, inorganic process (Smirnov, 1955).

In coals with high sulfur content, the sulfur occurs chiefly in pyrite, the oxidation of which, as in other ores, has been considered to be a process resulting from the chemical reaction of the pyrite with water containing dissolved oxygen. A consequence of this process is the formation of sulfuric acid.

The microbiological data now available compel us to assume that sulfides in the oxidation zone of ores and of coal are oxidized not only by inorganic but also by biogenic means, and the resulting increase in sulfate content in the water is one of the manifestations of biogenic transformation of the composition of water.

There is special interest in this connection in the autotrophic organism Thiobacillus ferrooxidans, which is able to oxidize chalcopyrite as well as pyrite (Ivanov, Lyalikova, and Kuznetsov, 1958; Lyalikova, 1959). An essential role in the formation of sulfate waters in the oxidation zone of ore deposits is played by sulfur-oxidizing bacteria (Thiobacillus thioparus and Thiobacillus thiooxidans), which are capable of oxidizing hydrogen sulfide, sulfur-oxygen compounds, and sulfides, and, probably, by autotrophic denitrifying bacteria that oxidize sulfur.

According to Kramarenko, in reference to the regions of rare-metal deposits of central Kazakhstan, sulfur-oxidizing bacteria are very widespread in unconsolidated sediments covering the ore bodies, obviously promoting an accumulation of sulfates in the ground water.

Pogrebitskii (1933), from his studies of sulfur in the coals of the Donets Basin, indicated that sulfur-oxidizing bacteria are the most likely cause of the variable sulfur content in coal beds. Subsequent microbiological investigations have confirmed the fundamental correctness of this hypothesis.

Ashmeed (1955), in describing the influence of microorganisms on the formation of sulfatic acid waters in a mine in Scotland, stated that the cause of the acidity is not only chemical, but also biogenic oxidation of pyrite, with the formation of iron hydroxide and sulfuric acid. For each ton of sulfuric acid formed by chemical oxidation, nearly four tons of sulfuric acid of biogenic origin is formed here. Thiobacillus ferrooxidans has been found in the water of the mine.

Experiments of Lyalikova (1959) have shown that, when chemical oxidation of pyrite that was being acted on by Thiobacillus ferrooxidans was almost completely absent, coal was freed from 25% of its contained sulfur in the course of one month.

The sulfate content in a vertical hydrogeological section of coal basins, such as the Donets Basin, decreases with depth (Shchegolev, 1948), and at great depths, such as in the Chelyabinsk graben, the water becomes practically sulfate free, though it is more highly mineralized. This change in composition, from the point of view of biochemical transformation of sulfur, is systematic.

In the upper part of the section, when molecular oxygen is present, the biogenic oxidation of pyrite is most intense; it becomes weaker with depth, where the optimum conditions for sulfate reduction are preserved.

It should be stated that biogenic formation of sulfates is related to a sulfate content, observed in a number of places, that is higher in the crestal zones of oil deposits than found in the water beyond the boundary of the oil accumulation. Two basic patterns have been observed: a) the ground water inside the oil-bearing zone is practically sulfate free, or it contains SO_4^{2-} in amounts considerably smaller than the water outside the zone; and b) the content of sulfates in the water at the margin of the oil zone is much greater than in the water beyond the boundary (Fig. 2).

The first type of sulfate distribution is to be found at a number of deposits in the Ural-Volga region, clearly seen in the Perm region adjoining the Urals (Krasnokamsk structure), and in some oil and gas structures of Fergana.

On the Apsheron Peninsula are deposits with sulfate distribution of both types. A classic example of water with higher sulfate contact in the crestal part of the structure and minimum sulfate in the zone beyond the oil boundary is found in the Surakhany structure, where purple sulfur bacteria were formerly found in the formational waters, these organisms gave the water a rose color (Isachenko, 1951). A similar distribution of sulfates is found at the Karachukhure deposit and in the Balakhany-Sabunchi-Ramaniny region. The presence of sulfate-free water at the margin of the oil zone has been noted at the Kala deposit.

The cause of a greater or lesser sulfate content in the crestal part of a structure may be associated in places with the inflow of deep ground water along faults, which are generally largest or most abundant in the arch of a structure. However, it is also possible, and we believe more likely, that this phenomenon may be explained by biogenic transformation of ground water.

Sulfate reduction is most intense in the zone immediately next to an oil deposit in the crestal part of a structure, in the ground water surrounding and underlying the deposit. This leads to a decrease in sulfate content of the marginal water as compared with water beyond the oil zone. These conditions correspond to the first of the indicated types of sulfate distribution in the ground water of oil deposits.

When the reduction of sulfates in the crestal part of a structure is most intense, conditions are developed that are most favorable for maximum accumulation of hydrogen sulfide in the ground water, and the environment is most suitable for oxidation of this constituent by sulfur, sulfur-oxidizing, and denitrifying bacteria— to sulfur and sulfuric

TABLE 2. Relationship between Content of Sulfates and Purple Bacteria in Water of the Surakhany Structure (after Volodin, 1935)

Well No.	SO_4, mg/liter	Color and position of the ground water relative to oil deposit
638	41	Colorless marginal water
699	291	
726	207	Intensely colored water in crestal part of structure at boundary of oil zone
506	303	
473	225	
578	96	Colorless marginal water

acid. The sulfuric acid that forms, reacting with salts in the ground water, chiefly carbonates, leads to an increase in sulfate content in the water within the oil zone; this corresponds to the second type of distribution of the SO_4^{2-} ion in ground waters of oil structures.

The sequence of changes in the amount of sulfate in the water, according to Volodin (1935), is clearly traced across the strike of the already mentioned Surakhany structure. The increased content of the SO_4^{2-} ion here corresponds to a more intense rose color of the water because of the pigment of the destroyed cells of purple sulfur bacteria of the genus Chromatium (Table 2).

From Table 2 it may be seen that the content of the sulfate ion decreases from the marginal and more distant parts of the deposit; a similar pattern may be observed along other cross sections of the same structure (Fig. 2).

The Biogenic Transformation of Compounds of Nitrogen and Other Elements in Ground Water

The activity of microorganisms in ground water is not restricted to the carbon and sulfur cycles, but also affects the migration of nitrogen. We shall mention some facts, pertinent to this point, which confirm the validity of identifying microorganisms with "chemical reagents" in ground water, as done by Omelyanskii.

Ammonifying bacteria, which decompose albumin compounds and, primarily, amino acids, are responsible for the biogenic generation of ammonia. It should be stated that ammonia may also be formed in the deep zones of the lithosphere at high pressures by purely chemical combination of nitrogen and hydrogen molecules.

The oxidation of ammonia by nitrifiers takes place chiefly in the upper hydrodynamic zone of the most intense water exchange, where molecular oxygen is present.

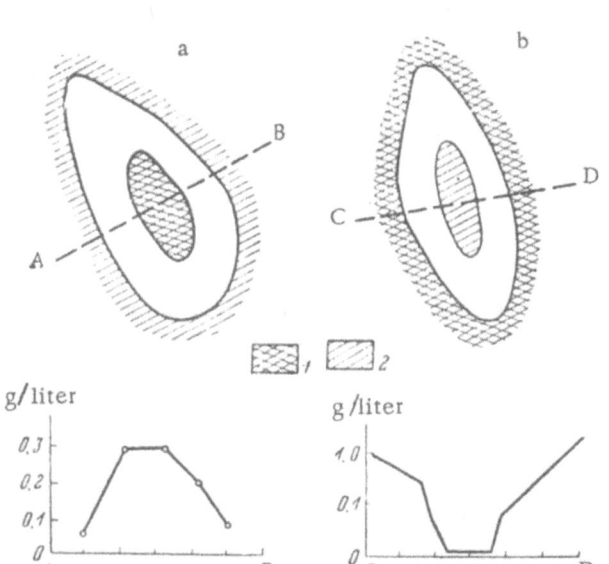

Fig. 2. Types of sulfate distribution in the ground water of oil and gas structures: a) change in sulfate content of water across (along A-B) the strike of the Surakhany deposit (after N. A. Volodin); b) change in sulfate content along the strike (C-D) of the Yuzhni Alamyshek deposit in Fergana (after B. B. Mitgarts); 1) ground water with a high content of the SO_4^{2-} ion; 2) ground water with a low content of the SO_4^{2-} ion and water practically sulfate free.

In the zones of more difficult and very difficult water exchange, nitrification is sharply curtailed and nitrates are reduced by denitrifying bacteria to free nitrogen.

In the vertical hydrogeologic section of artesian basins, one may observe a certain zonal arrangement in the biogenic transformation of nitrogen compounds, due to the nature of water exchange and to the position of the oxygen boundary. The presence of nitrates in deep artesian waters is insufficiently clear and requires further study.

Something else that is not altogether clear is the role of bacteria in the geochemistry of ground water as a factor governing the transfer of phosphates into solution by biogenic synthesis and controlling the decomposition of phosphates. There has been little evaluation, applicable to the development of the composition of water, of the importance of iron bacteria, which oxidize large quantities of ferrous oxide, converting it into insoluble hydroxide.

The Ecological-Geochemical Pattern of Distribution of Microorganisms in Ground Water

The examined data lead one to conclude that almost all ground water in sedimentary rocks and, apparently, to a certain extent, in volcanic rocks as well, has been subjected to biogenic transformation of its composition over the extent of geologic history. Beginning with the stage of early diagenesis in the sediments (muddy waters) and ending with the stage of late katagenesis of rocks, microorganisms preserve their biogeochemical importance as a factor influencing the regrouping of atoms in the composition of the water.

Even the scanty data cited above, concerning the role of bacteria in the geochemistry of ground water at metallic and nonmetallic deposits, demonstrate the significance of bacteria in the aqueous migration of chemical elements and of biocatalysts in a number of hydrogeochemical processes.

Microorganisms are adapted to the temperature, degree of mineralization, and composition of the water; and the total number in and the nature of the biocoenosis depend on the rate of water exchange in the ground. In this way ecological-geochemical patterns are developed among the microorganisms in soil and artesian water. One of the essential conditions of this distribution is the adaptability of the microorganisms to the temperature of the ground water, leading to the development of thermophilic strains. A zonal pattern of distribution of microorganisms is noted with depth for the USSR, conditioned by the geothermal conditions inside the earth. This zonal pattern, in general plan, is related to the climatic zonal pattern of artesian waters in the USSR, and it involves the fact that the temperature of optimum development of bacteria increases regionally from north to south (Gurevich, 1958a). Ground waters, possessing a common content of one or several chemical elements that give rise to the biological reaction of the microorganisms living in the waters, may be separated into biogeochemical zones or provinces. This means of studying biogeochemical patterns, in its indicated principal features by A. P. Vinogradov (1949), makes it possible to display the mutual relations between natural waters and microorganisms.

As an example we may mention the biogeochemical zone of water in oil deposits, which is characterized by virtual freedom from sulfate and by the presence of sulfate-reducing bacteria. The biogeochemical zone of mine waters is characterized by a high content of sulfate and by the presence of bacteria capable of oxidizing the sulfur in sulfides. Correspondingly there are biogeochemical zones in muddy waters, ground waters, oxidation zones of ore deposits, and elsewhere.

The study of these and other biogeochemical zones in ground water, with due consideration to the actual geologic and hydrogeologic conditions under which they are found, requires the combined forces of microbiology and hydrogeology. The principal task is a quantitative appraisal of the effectiveness of microbiological transformation of the chemical composition of water. Only in this way can we evaluate the geochemical significance of this process.

Conclusions

1. The essential but inadequately studied factor of geochemistry of ground water is biogenic transformation, i.e., changes in the ionic and gaseous composition through the activity of microorganisms.

2. The condition by which bacteria appear in ground water at depths of thousands of meters have been inadequately investigated. The hypothesis of a sedimentational origin of the microflora ties this phenomenon to the inheritance of muddy waters at the stage of early diagenesis of sediments. During subsequent gravitational compaction of predominantly clayey deposits, most of the bacteria, together with expressed water, migrate into water-

bearing horizons, producing a bacterial population. This does not exclude the penetration of bacteria from the surface with infiltration waters during continental interruptions of the sedimentation cycle. The idea of a "sterile interior" and of introduction of bacteria into deep waters only during the drilling of boreholes in recent time loses its significance.

3. Almost all ground water in the biosphere has undergone biogenic alteration of its chemical composition in some degree and at various stages in its geologic history.

Microorganisms are adapted to the physical and chemical conditions of the underground hydrosphere; this means, primarily, adaptation to the temperature, mineralization, and composition of the water. In this way a spatial pattern of ecological-geographic distribution of microflora develops in soils and artesian waters. At the same time the number of microorganisms and the character of their biocoenosis depend on the intensity of underground water exchange, including artificially induced exchange such as occurs during exploitation of an oil deposit.

The climatic zonal pattern of artesian water described by N. I. Tolstikhin matches the regional zonal adaptation of microorganisms to the optimum temperature of their development.

The increase in temperature and mineralization of water with depth limits and, in the end result, precludes bacterial life at great depths.

4. Discovery of the ecological distribution of microflora with depth and in areal relationship to water-bearing rocks is of practical importance in the search for mineral deposits. The biogeochemical zone of ground water is the principal feature reflecting this kind of systematic spatial distribution. In agreement with the principle of A. P. Vinogradov, this zone is distinguished by the similarity of one or several indicators of the chemical composition of the water, produced by the corresponding biological reaction of the microflora. Individual biogeochemical zones are defined by definite characteristics of biogenic transformation, such as the presence of sulfate-free, hydrogen-sulfide, ferruginous, and other waters.

5. The biochemical transformation of sulfur and carbon, directly connected with the migration of these elements, introduces profound changes in the ionic and gaseous composition of ground water. Data from Soviet investigators indicate a similarity in the changes in composition of muddy waters in seas and oceans and of ground water through the reduction of sulfates; these data furnish grounds for the hypothesis of biogenic formation of sodium bicarbonate alkaline waters, which characterize a number of oil deposits.

6. The inverse relationship between sulfate content of water and the oxidizability of adjacent oil, traced through the geochemical sections of oil basins, is obscured and complicated by the ancient processes of biochemical transformation of sulfur and carbon. Relict traces of these transformations in the composition of the water, the oil, and, apparently, the physiology of the microorganisms are associated in many places with formations that were exposed to continental interruptions of the sedimentation cycle (Apsheron and Volga-Ural oil and gas regions).

7. Along with the reduction of sulfates, another hydrogeochemical feature of the sulfur cycle in ground water should occupy our attention— the biogenic reduction of sulfates during bacterial oxidation of hydrogen sulfide, molecular sulfur, and metallic sulfides. This source of secondary SO_4^{2-} ions is clearly of great quantitative significance in the ground water about deposits of oil, rock salt, and sulfide ore. For prospecting, an important factor is the activity of sulfate-reducing bacteria in the formation of geographic ranges of waters with high sulfate content at sites of sulfide mineralization and in the crestal zones of oil structures.

8. The biochemical transformation of nitrogen in artesian basin conforms to a vertical zonal pattern, determined chiefly by the depth to the oxygen boundary. The origin of small traces of nitrites and nitrates, which are found in artesian waters even at great depths, is not altogether clear.

9. A clarification of the conditions of microorganisms adaptation to the geochemical and hydrogeological environment is the basis of scientifically grounded appraisal of the value of the microorganisms in prospecting and of their role in the aqueous migration of chemical elements. It is necessary to expand our efforts in this direction by the systematic joint work of hydrogeologists, microbiologists, and chemists.

LITERATURE CITED

Alekin, O. A. 1953. Fundamentals of Hydrochemistry [in Russian], GIMIZ.
Al'tovskii, M. E., Kuznetsova, Z. I., and Shvets, V. M. 1958. The Origin of Oil and Oil Deposits [in Russian] (available in English translation from Consultants Bureau Enterprises, Inc.).

Ashmeed, D. 1955. "The influence of bacteria in the formation of acid mine water." Colliery guard., 190, p. 694.

Beder, B. A. 1958. "Water of artesian basins." Uzbekistanskii geologicheskii zhurnal, No. 6.

Belyakov, E. A. 1956. "The oil-prospecting value of ground water and of gases dissolved in the water, from data on investigations in the Sakhara (Samara?)-Kama interstream region of the Volga-Ural oil region." from the Collection: Questions on Oil-Prospecting Hydrogeology [in Russian], Gosgeolizdat.

Dutova, E. N. 1959. "The oil-prospecting value of microflora in the ground waters of some regions in the western part of Central Asia." Ground Water [in Russian], Inform. sb. VSEGEI, No. 19, Leningrad.

Grigor'ev, S. M. 1954. The Formation and Properties of Fossil Fuels [in Russian], Izd. AN SSSR.

Gurevich, M. S. 1954. "Data on oil-prospecting hydrogeology in the southern part of the Western Siberian Lowland." All-Union Geological Institute: Data on the Geology, Hydrogeology, and Oil and Gas Possibilities of Western Siberia [in Russian], General Series, No. 1, Gosgeoltekhizdat.

Gurevich, M. S. 1958a. "Biogeochemical transformation of sulfur in ground water at oil and gas deposits." Summary Reports of the Twelfth Hydrochemical Conference on Studies of Chemical Processes Taking Place in Natural Waters [in Russian], Gidrokhimicheskii institut AN SSSR, Novocherkassk.

Gurevich, M. S. 1958b. "Some factors in the biogenic transformation of ground water." Trudy Lab. gidrogeol. problem im. akad. F. P. Savarenskogo, 16.

Isachenko, B. L. 1951. "Purple sulfur bacteria from the lower boundary of the biosphere." Collected works, 2 [in Russian], Izd. AN SSSR.

Ivanov, M. V., Lyalikova, N. P., and Kuznetsov, S. I. 1958. "The role of sulfur-oxidizing bacteria in the weathering of rocks and sulfate ores." Izv. AN SSSR, seriya biol., No. 2.

Kolesnik, E. A. 1951. "Microflora in water and oil of the Vtoroi Baku region." Trudy VNIGRI, novaya seriya, No. 57.

Kuznetsov, S. I. 1952. The Role of Microorganisms in the Cycle of Substances in Lakes [in Russian], Izd. AN SSSR.

Lindtrop, N. T. 1947. "The reduction of sulfates in the Grozny oil fields." Doklady Akad. Nauk SSSR, 57, No. 9.

Lyalikova, N. N. 1959. Author's abstract of her dissertation: Physiology and Ecology of Thiobacillus ferrooxidans in Connection with Its Role in the Oxidation of Sulfide Ores [in Russian].

Malyshek, V. T. and Shaikhet, P. A. 1959. "The saline composition of sediment solutions (muddy waters) in the bottom deposits of the Caspian Sea, as compared with the composition of aqueous extracts from the same sediments." Trudy Azerb. issl. inst. po dobyche nefti. No. 8, Baku.

Mitgarts, B. B. 1956. "The oil-prospecting significance of the composition of ground water, from data on investigations in Fergana." from the Collection: Questions on Oil-Prospecting Hydrogeology [in Russian], Gosgeolizdat.

Perfil'ev, B. V. 1932. "Biology of medicinal muds." from the Collection: Principles of Health-Resort Practice [in Russian] 1, Gosmedizdat.

Pogrebitskii, E. O. 1933. "Sulfur in the coals of the Donets Basin." Problemy geologii, No. 5.

Shchegolev, D. I. 1948. Mine Waters [in Russian], Ugletekhizdat.

Shishkina, O. V. 1958. "The saline composition of muddy waters of the Far Eastern seas and adjacent parts of the Pacific Ocean." Problems of Marine Chemistry [in Russian], Trudy Inst. okeanologii, 26.

Shiskina, O. V. 1959a. "Chemical composition of muddy waters of the Pacific Ocean." Problems of Marine Chemistry [in Russian], Trudy Inst. okeanologii, 33.

Shishkina, O. V. 1959b. "Sulfates in muddy waters of the Black Sea." Problems of Marine Chemistry [in Russian], Trudy Inst. okeanologii, 33.

Smirnov, S. S. 1955. The Oxidation of Sulfide Deposits, 3rd Ed. [in Russian], Izd. AN SSSR.

Stadnikov, G. L. 1957. Argillaceous Rocks [in Russian], Izd. AN SSSR.

Strakhov, N. M. 1951. "The limestone-dolomite facies of recent and ancient sedimentary basins," Trudy Inst. geol. nauk, No. 124, geol. seriya (No. 45).

Sulin, V. A. 1948. The Hydrogeology of Oil Deposits [in Russian], Gostoptekhizdat.

Tageeva, N. V. 1955. "Experimental investigations on the study of the origin of alkaline-earth sodium chloride formational brines." Problems of Studying Ground Water and Engineering-Geological Processes [in Russian], Izd. AN SSSR.

Tageeva, N. V. 1958. "Principal geochemical types of ground water." Trudy Lab. gidrogeol. problem im. akad. F. P. Savarenskogo, 16.

Tageeva, N. V. and Tikhomirova, M. M. 1957. "Some features of early diagenesis of sediments in the northwestern part of the Black Sea." Doklady Akad. Nauk SSSR 112, No. 3.

Vernadskii, V. I. 1935. Problems of Biogeochemistry I: The Value of Biogeochemistry for Probing the Biosphere [in Russian], Izd. AN SSSR.

Vinogradov, A. P. 1949. "Biogeochemical provinces." Proceedings of the Jubilee Session of the Academy of Sciences, USSR, Commemorating the Hundredth Anniversary of V. V. Dokuchaev's Birth [in Russian], Izd. AN SSSR.

Volodin, N. A. 1935. "Rose-colored borehole waters of the Ordzhonikidze Industry." Azerb. neft. khoz-vo, No. 6.

Zhabref, D. V. and Khachkevich, N. I. 1951. "Development of water at oil deposits." Neftyanoe khoz-vo, No. 12.

THE SIGNIFICANCE OF SULFATE-REDUCING BACTERIA
IN PROSPECTING FOR OIL, AS EXEMPLIFIED IN THE STUDY
OF GROUND WATER IN CENTRAL ASIA

E. N. Dutova

(All-Union Scientific-Research Institute of Geology, Leningrad)

A study of the microflora in ground water of Central Asia has led to joint prospecting for oil by the Hydrogeological Branch and the Central Asian Expedition of the All-Union Geological Scientific-Research Institute (VSEGEI). This is a continuation of regional studies in the regions of Vtoroi Baku, Fergana, and the Western Siberian lowland. The principal purpose of the microbiological investigations is the clarification of the character and scale of biogeochemical transformations in ground water due to microflora; another objective is the appraisal of the significance of individual groups of bacteria and their biocoenoses in prospecting for oil. In recent years the ground water of oil-producing horizons in Central Asia has been studied in the Turkmen SSR and Uzbek SSR (the Kokaity, Lyal'-Mikar, Khaudag, and Nebit-Dag deposits) and in the Bukhara-Karshi region, where exploratory work is going on the present time. For comparison, ground water has been studied from zones not associated with oil deposits, from the northern foothills of the Kopet-Dag, from the Bolshoi Balkhan, the Krasnovodsk Peninsula, and other places. The samples of water were collected during hydrogeological tests on flowing wells and also from pumped wells. The microbiological analyses have shown that the microflora of ground water associated with oil deposits consists chiefly of groups of bacteria participating in the transformation of sulfur: Thiobacillus thioparus, Th. thiooxidans, Th. denitrificans and sulfate-reducing bacteria. About 40% of the investigated samples of ground water contain bacteria that destroy naphthenic acids; rarely are bacteria encountered that ferment glucose with the formation of gas; and anaerobic cellulose bacteria are missing entirely. The total number of saprophytic, denitrifying, and ammonifying bacteria fluctuates according to the degree the structure is a closed system: from individuals and tens of cells per milliliter of ground water in closed structures to tens and hundreds of thousands of cells in the same quantity of water that circulates in beds exposed at the surface (in places) or in a zone of free water exchange. In several samples of ground water it was impossible to find one of the 13 definite physiological groups of bacteria. This barrenness is generally associated with low or high values of pH, high temperatures, and strong mineralization of the water. A basic distinction of the microflora in ground water in contact with petroleum hydrocarbons in nature is the wide distribution and the high activity of sulfate-reducing bacteria in such water. In an overwhelming number of experiments with inoculations of ground water from oil and gas-oil deposits, sulfate reduction was observed, with utilization not only of organic acids but of higher hydrocarbons of the paraffin series as well: heptane, octane, and nonane.

In inoculations with ground water not associated with oil deposits, this process was not once observed. The presence of sulfate-reducing bacteria growing on heptane and nonane in the formational waters of oil deposits was previously noted by Kuznetsov (1950) and Kramerenko (1956). Consequently, the presence of sulfate-reducing bacteria growing on nonane and hydrocarbons of lower molecular weight is characteristic of water associated with oil deposits. However, as a study of ground water from oil fields in Central Asia has shown, about 30% of the investigated samples exhibited no reduction of sulfates in media with hydrocarbons; the process occurred only in media with organic acids. Sulfate reduction by utilization of hydrocarbons is mostly observed only in inoculations of water that contain, besides sulfate-reducing bacteria, ammonifying, denitrifying, and other groups of bacteria that grow on other ordinary culture media. The more numerous and variable the microflora in formational waters, the more active the process of sulfate reduction to hydrocarbons becomes, and the simpler the hydrocarbons that arise (down to and including hexane). In order to explain this phenomenon and to evaluate the significance of sulfate-reducing bacteria in prospecting for oil, more detailed studies were made on this group. These experiments have shown that the capacity to use the hydrocarbons heptane, octane, and nonane in the process of reducing sulfates is inherent in bacteria extracted from ground water in contact with petroleum hydrocarbons. Sulfate-reducing bacteria extracted from water not in contact with petroleum hydrocarbons do not possess this capacity.

Sulfate-reducing bacteria grown on hydrocarbons in the laboratory may change and use other organic substances (calcium lactate). During subsequent re-inoculations of these bacteria, the capacity to use hydrocarbons is preserved. It is obvious that this capacity is acquired by the sulfate-reducing bacteria during long contact with hydrocarbons under natural conditions. A specific use of hydrocarbons has not been observed: a particular culture of bacteria may be grown on nonane, octane, or heptane. It is easiest to use nonane, the hydrocarbon with the greatest molecular weight.

It has been shown by a series of experiments that the use of hydrocarbons (nonane, octane, and heptane) during the reduction of sulfates takes place only in a composite culture of sulfate-reducing bacteria. In purifying a culture from accompanying microflora, the capacity to grow on hydrocarbons of molecular weight lower than decane is lost. Pure cultures of sulfate-reducing bacteria extracted from individual colonies have been found unable to cause sulfate reduction with the use of hydrocarbons. When meat-peptone agar is inoculated with composite cultures of sulfate-reducing bacteria, individual, characteristic, wrinkled and smooth colonies of bacteria appear, accompanying the principal culture. These accompanying bacteria, being small mobile bacilli, are found to be able to reduce nitrates to gaseous nitrogen, to ferment carbohydrates, and to oxidize hydrocarbons in an aerobic environment (heptane-nonane). From preliminary data these may be assigned to the genus Pseudomonas.

The introduction of individual colonies of these bacteria into a pure culture of sulfate-reducing bacteria almost always leads to sulfate reduction on Tauson and Veselov media with hydrocarbons. The experiments that were conducted permit us to explain the absence of sulfate reduction in media with hydrocarbons inoculated with some samples of formational water from the oil fields of Central Asia. The waters in the samples were greatly impoverished in microflora, and the sulfate-reducing bacteria living in them apparently were not accompanied by sufficiently active microflora. When easily available nutrient substances are introduced into such impoverished water, substances promoting the development of the microflora in the water, reduction of sulfates is observed on nonane and heptane. The problem of the role of these accompanying microflora in the reduction of hydrocarbons and the reduction of sulfates requires further study. It might be assumed that the accompanying bacteria first destroy all the hydrocarbon molecules, supplying more highly oxidized hydrocarbons for feeding the sulfate-reducing bacteria. It is possible that the oxidation-reduction potential is lowered through the life activity of these organisms, bringing the value to the optimum sulfate-reducing bacteria. The data we have obtained have been confirmed by the work of Simakova and others (1958), who have shown that in the first stages of oil transformation, it is not sulfate-reducing bacteria that grow, but other groups, particularly Pseudomonas fluorescens denitrificans. The Czechoslovakian investigator Dostalek (Dostalek and Spurny, 1958) has also noted in his work that the oil yield of a bed is increased by the joint introduction into the oil deposit of sulfate-reducing bacteria and bacteria from the genus Pseudomonas.

The cited material leads us to conclude that sulfate reduction by utilization of hydrocarbons of lower molecular weight than decane is effected not by a single physiological group, but by a complex of bacteria inhabiting ground water in contact with petroleum hydrocarbons.

The process may be recommended as an indicator in oil prospecting only when the ground water contains abundant microflora and the conditions are suitable for their development. When the ground water is very poor in microflora, supplementary microbiological investigations must be made.

Conclusions

1. It has been shown that the reduction of sulfates in media with heptane, nonane, or octane takes place only in cultures of composite sulfate-reducing bacteria that have been taken from water in contact with petroleum hydrocarbons.

2. In separating cultures of sulfate-reducing bacteria from accompanying bacteria, the first lose the capacity to grow on hydrocarbons of lower molecular weight than decane.

3. Most of the accompanying bacteria are assigned to the genus Pseudomonas.

4. Sulfate-reducing bacteria may be recommended as indicator organisms in prospecting for oil only when the ground water teems with microflora.

LITERATURE CITED

Dostalek, M. and Spurny, M. 1958. "Bacterial release of oil." Folia biol. No. 3, pp. 166-172.

Kramarenko, L. A. 1956. "The composition and distribution of microorganisms in ground water and their value in prospecting." from Questions on Oil-Prospecting Hydrogeology [in Russian].

Kuznetsov, S. I. 1950. "A study of the possibility of present-day formation of methane in the gas and oil facies of the Saratov and Buguruslan regions." Mikrobiologiya, 19, No. 3.

Simakova, T. L., Gorskaya, A. I., Kolesnik, Z. A., and others. 1958. "Nature of changes in oil in anaerobic environments due to the biogenic factor." Trudy VNIGRI, No. 128.

BIOGENIC REDUCTION OF SULFATES IN FORMATIONS
DURING FLOODING OF OIL DEPOSITS BY SEA WATER

M. V. Gasanov

(Azerbaidzhan Scientific-Research Institute for Oil Production, Baku)

As a result of pumping sea water or mixtures of sea water and other water into the beds of the Apsheron deposits, microbiological reduction of sulfates has been observed in the formations.

As a consequence, huge quantities of hydrogen sulfide have been formed. The presence of this constituent in the formational water causes increased corrosion of the steel pipes of the deep pumps and corrosion damage to the rods. All this leads to frequent interruptions in operation of the well, and is strongly reflected in the yield of oil.

Therefore, a study of hydrogen-sulfide fermentation in oil beds becomes of great practical significance for the national economy.

Conditions in the oil beds of the upper and lower divisions of the productive sequence (to a certain depth) are fully suitable for the growth of sulfate-reducing bacteria.

But the absence of a sufficient quantity of sulfates in the formational water is strongly reflected in the activity of the sulfate-reducing bacteria (ZoBell, 1958). It requires only the artificial introduction of the SO_4^- ion into the oil bed to activate the sulfate-reducing bacteria. Therefore, the cause of hydrogen-sulfide generation in formational water that contains no SO_4^- ions is the artificial introduction of SO_4^- ions with sea water during attempts at secondary oil recovery. In 1958 we made studies on the appearance of hydrogen sulfide at the First and Thirteenth Companies of the Petroleum Industry Administration (NPU) "Ordzhonikidzeneft'" and at the Third Company of the NPU "Artemneft'" (Gasanov and Akhundov, 1959).

The results of these studies have shown that when parts of horizons II and III at "Ordzhonikidzeneft'" were flooded with sea water (the beds previously contained nothing but formational water), the reduction of sulfates by microorganisms became intense in the parts of the beds between the input well and the first row of producing wells. In this process the amount of hydrogen sulfide form amounted to as much as 600-650 mg/ liter in individual samples (see figure).

Further studies on the fermentation of hydrogen sulfide were made in the Kosha-Naur area, in horizon V of the Balakhany series (First Company of NPU "Leninneft'").

The Kosha-Naur area began to be developed in 1934. In 1951 gas pumping was started in the area, and by March of 1954 waterflooding operations were commenced. The flooding was done first with water from Lake Zabrat, but later with a mixture of three types of water in equal proportions: from the sea, Beyuk-Shor (salt marsh), and Lake Zabrat.

The appearance of hydrogen sulfide in water of the producing wells of horizon V at the First Company was noted only three years (1957) after flooding began.

Twelve samples of water were collected and analyzed for the eastern field. For the samples first collected the change in H_2S content ranged from 8.5 to 345 mg/ liter. The total amount of H_2S in this field increased to such an extent that in some wells the quantity of H_2S in mg/ liter in the second sample, collected two months after the first, was several times the amount in the first sample. In well 1018 the amount of H_2S after $2\frac{1}{2}$ months increased

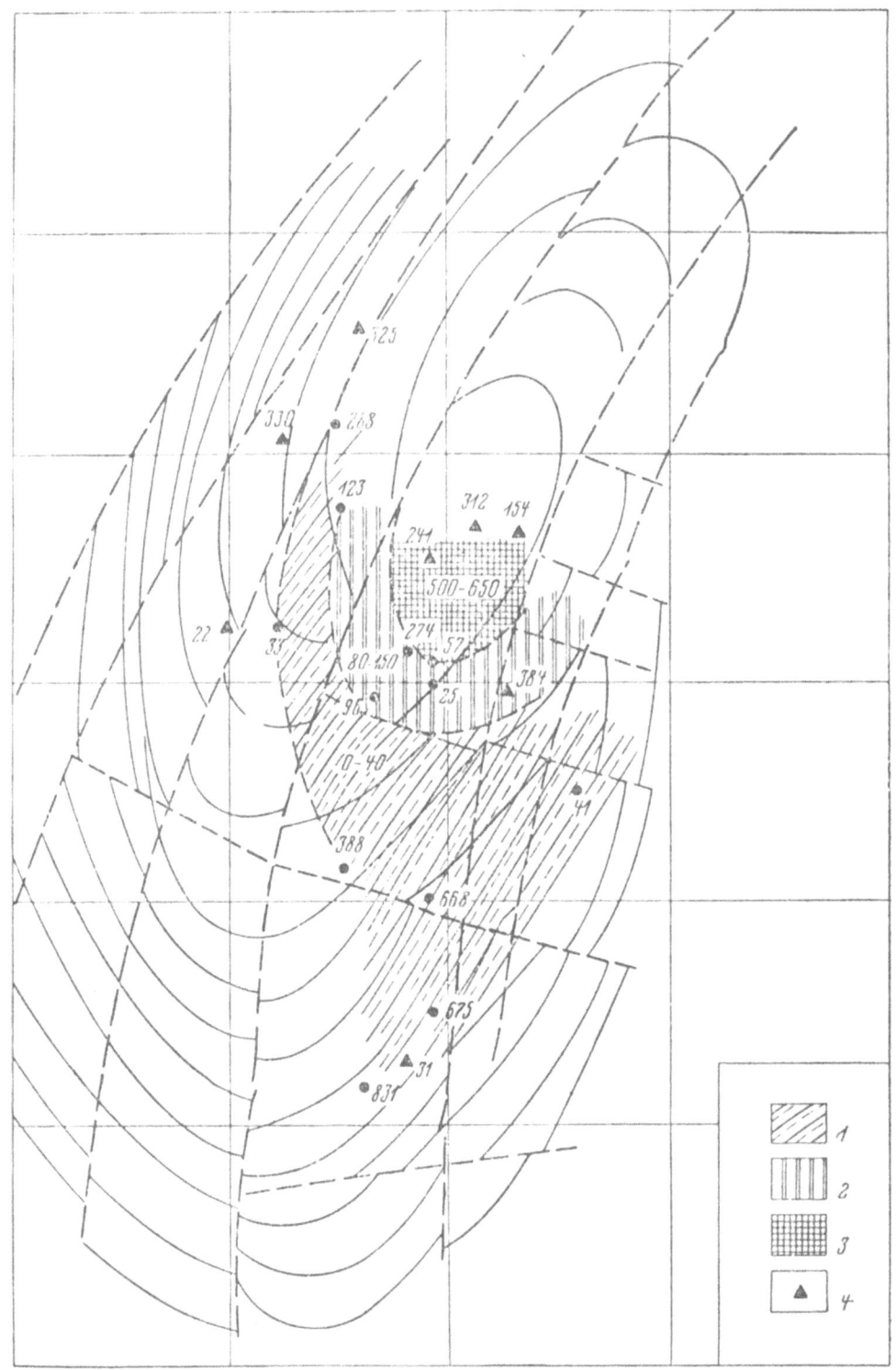

Map showing distribution of hydrogen sulfide in the third horizon of the "Ordzhonikidzeneft'"
Company: 1) content of hydrogen sulfide 0-40 mg/liter; 2) content of hydrogen sulfide
80-150 mg/liter; 3) content of hydrogen sulfide 500-650 mg/liter; 4) injection wells.

to approximately eight times the previous quantity (the first determination gave 19.81 mg/liter, the second 163.5).
An increase in quantity of H₂S was observed in the water of all 12 investigated wells of the indicated field (Table 1).

All this graphically points to intense biogenic reduction of sulfates, which formed by pumping water containing SO₄″ ions into the bed.

TABLE 1. Number of Sulfate-Reducing Bacteria and Content of H$_2$S for Various Periods of Observation. Kosha-Naur Deposit

Well No.	Date of first sample	No. of bacteria per ml of water	Amount of H$_2$S, mg/liter	Date of second sample	No. of bacteria per ml of water	Amount of H$_2$S, mg/liter
1173	Mar. 9	90	125.7	May 25	115	142.7
1020	Mar. 16	108	102.89	June 3	110	178.78
1018	The same	24	19.81	The same	130	163.46
931	" "	23	38.79	" "	120	89.69
1054	" "	50	31.80	" "	100	73.71
1008	April 1	110	211.16	" "	150	257.4
1049	The same	56	42.48	June 20	95	75.0
1067	" "	40	39.1	The same	100	100.86
1040	April 8	69	8.5	June 10	75	38.8
1064	The same	120	11.9	The same	126	32.0
1060	April 15	140	69.10	June 20	150	88.20
1185	The same	110	8.28	June 22	150	61.78
1086	April 15	72	173.3	June 30	90	252.72
1094	The same	125	45.32	The same	147	122.04
2466	The same	60	264.9	" "	70	273.0
1069	" "	117	104.6	" "	120	120.68
2783	May 11	130	163.37	July 11	163	175.77
2053	The same	50	95.50	The same	120	154.92
2476	" "	130	37.29	" "	200	85.09
1405	" "	50	14.56	Aug. 15	110	59.33

The area of distribution of the central field was greater than either the eastern or western alone. The greatest number of water samples (16) were collected and analyzed for this field. In order to trace the changes in H$_2$S content with time for all these wells, samples of water were collected after definite intervals of time.

The chemical analyses of water samples collected repeatedly for a given field show that the content of hydrogen sulfide increased considerably. Observations were made on five wells for the western field. On April 11, 1959, the amount of H$_2$S in mg/liter ranged from 14.6 to 163.4 mg/liter. Samples of water were again collected from these wells after an interval of 2-2.5 months. Analyses showed an increase in H$_2$S in all wells. The H$_2$S content in these second samples ranged from 46.9 to 175.8 mg/liter. In individual wells (2476 and 1405), the H$_2$S content, despite a short interval between sampling, increased twofold and even threefold.

A Study of the Change in Quantity of H$_2$S in Relation to the Change in Number of Sulfate-Reducing Bacteria

Since sulfate-reducing bacteria play a very important role in the formation of hydrogen sulfide, both in sulfur springs and in formational water, it became a matter of interest to study the effect of the number of sulfate-reducing bacteria on the content of hydrogen sulfide in formational water during flooding of the formation with sea water. To investigate this relationship, computations were made on sulfate-reducing bacteria and on the content of hydrogen sulfide in the first and second (after 2-2.5 months) samples of water collected from producing wells in the Kosha-Naur area.

For the computations 10-ml test tubes were used, 5 mm in diameter and 350 mm long. It was possible, without great error, to count the number of colonies of sulfate-bacteria in these test tubes, growing on agar-thickened Van Delden medium. For these experiments agar-thickened Van Delden medium and a medium of mixed natural waters (Gasanov and Akhundov, 1959) were used; the inoculation was made by successive dilutions by the power of 10 according to the well-known method. The counts were generally made on the 24th day after inoculation. From the data in the table it may be seen that the number of sulfate-reducing bacteria increased in all samples of water from the producing wells.

A similar change is observed in the content of hydrogen sulfide in the same samples.

As an example we may note the results from the producing well 1018; in one milliliter of water of a sample collected on March 16, 1959, 24 sulfate-reducing bacteria were counted and the hydrogen-sulfide content was 19.81 mg/liter. A study of the second sample of water, collected from the same well 77 days later, shows an increase in the number of sulfate-reducing bacteria to 130 per milliliter of water and a rise in quantity of hydrogen sulfide to 163.5 mg/liter.

After discovery of the principal cause of hydrogen-sulfide formation, an attempt was made to prevent hydrogen-sulfide fermentation. For this purpose we tested some bactericides, the most effective of which is formaldehyde.

Experiments with formaldehyde have shown that the growth of sulfate-reducing bacteria is suspended when formaldehyde is introduced in quantities of 40-100 mg/liter.

As laboratory experiments have shown, with this concentration of formaldehyde, nutrient Van Delden medium and a mixture of natural waters (50% sea water and 50% alkaline water) stood a long time with no change; no formation of H_2S was observed. In a control experiment with no formaldehyde, sulfate reduction began on the third day of the experiment. After these experiments we recommended that formaldehyde (100 mg/liter) be pumped into the space between pipe and wall of producing wells at the first NPU Company "Leninneft'."

Changes in the Number of Sulfate-Reducing Bacteria during Injection of Formaldehyde

After formaldehyde had been pumped into the wells for 1-2 days, samples were collected and determinations were made on the H_2S content and the number of sulfate-reducing bacteria.

The results of calculating the sulfate-reducing bacteria are shown in Table 2.

TABLE 2. Number of Sulfate-Reducing Bacteria in Formational Water Before and After Pumping Formaldehyde into the Space between Pipe and Wall

Well No.	No. of bacteria per ml of water		Well No.	No. of bacteria per ml of water	
	before pumping	after pumping		before pumping	after pumping
1025	80	30	1094	147	None
1008	150	15	1095	87	14
1018	130	28	925	235	25
1020	110	26	931	120	45
1057	34	None	1045	50	None
1048	200	15	1054	100	30
1048	95	53	1405	110	20
1060	150	None	2466	80	None

As seen in Table 2, after the formaldehyde had been pumped in, the number of bacteria in the water samples dropped sharply, in some samples no bacteria being found at all; the H_2S content in the samples of water also decreased. This phenomenon may be associated with the view that part of the hydrogen sulfide forms in the space between pipe and wall, where sulfate-reducing bacteria live on the inner surface of the pipe and there forms H_2S by hydrogenous reduction of sulfate. Therefore, after formaldehyde has been pumped in, the activity of these bacteria is suspended and the formation of hydrogen sulfide ceases.

Conclusions

1. The water in wells producing from horizon V in the Kosha-Naur area is thoroughly contaminated with hydrogen sulfide.

2. In a given bed, the formation of H_2S is governed by biogenic reduction of sulfates; this process is stimulated by sulfate-reducing bacteria.

3. Hydrogen-sulfide fermentation in the Kosha-Naur area developed after flooding horizon V.

4. It has been ascertained that the amount of hydrogen sulfide in a producing well increases with increase in number of sulfate-reducing bacteria in the water.

5. The greatest number of breaks in the rods have been observed in holes where the content of H_2S is high.

6. It has been determined experimentally that formaldehyde in the amount of 100 mg/liter prevents the growth of sulfate-reducing bacteria.

7. After formaldehyde has been pumped into the space between pipe and wall the number of sulfate-reducing bacteria in the water is diminished sharply.

LITERATURE CITED

Gasanov, M. V. and Akhundov, A. R. 1959. "The formation of hydrogen sulfide in formational waters of the Kosha-Naur petroleum industry." Azerb. neft. khoz-vo, No. 5.

Malyshek, V. T. and Gasanov, M. V. 1959. "A study of sulfate reduction in mixtures of sea water and alkaline formational water." Trudy Azerb. n.-i. inst. po dobyche nefti, No. 8.

ZoBell, C. E. 1958. "Ecology of sulfate-reducing bacteria." Producers Monthly, 22, No. 7.

LIFE ACTIVITY OF FORMATIONAL MICROFLORA
AS AN INDEX OF GEOLOGIC ENVIRONMENT AND PROCESSES
OBTAINING IN PETROLIFEROUS FORMATIONS

K. B. Ashirov

(State Institute for Planning and Study of the Oil-Production Industry, Kuibyshev)

In developing the oil fields of the middle Volga region, data have continued to accumulate in confirmation of an active role of microorganisms in the over-all cycle of geologic processes.

The work of a great number of microbiologists (Kuznetsov, 1957; Shturm, 1941, 1950a, 1950b; Éksertsev, 1951; Kolesnik, 1955; Kuznetsova and others, 1957) has established the fact that a rich and varied microflora is associated with oil beds. This fact indicates that biogenic processes have a definite place in the general chain of biogeochemical transformations in sedimentary rocks, the results of which we encounter in our everyday activity.

At the present time we may consider the following to be established facts concerning the basic pattern of distribution of formational microflora and the influence of these microflora on the oil deposits of the middle Volga region.

1. Microflora are found in the entire section of sedimentary rocks of Paleozoic age, to depths of 3000 m and below, where, as an example, formational pressure reaches 350 atm and the temperature rises to 74° in bed D-IV of the Mukhanovo deposit.

2. The microflora has a wide regional distribution and is found within the earth at practically all the oil deposits of the Volga region.

3. It has been ascertained that the microflora occurs chiefly in the contact zone between oil and formational water.

4. Microflora lives in deposits with various kinds of oil, in waters of various types and different degrees of mineralization, and also in different beds (reservoirs).

5. The greatest development of microflora is found in the upper Paleozoic section. Desulfurizing bacteria are practically restricted to this part of the section of the investigated region. The growth of microflora is slight and desulfurizing forms are practically absent in clastic Devonian strata.

6. It has been discovered that the number of microflora increase with time during the development of oil deposits (observed in a number of places).

7. It has been ascertained that the destruction of oil deposits and the generation of biogenic gases (H_2S, CO_2, and others) are associated with the life activity of formational microflora.

On the basis of patent fundamental patterns, it is possible to trace in concrete examples the influence of the geological environment on the trend of biochemical processes in oil beds and the effect of these processes on the geological environment within the earth.

The relationship between formational microflora and oil deposits is confirmed by the fact that rocks overlying and underlying oil-impregnated beds are generally barren of microorganisms, whereas they are present in the oil beds (Ékzertsev and Kuznetsov, 1954).

In studying the growth of bacteria in oil deposits, it has been discovered that the abundance of the bacteria depends on the degree of flooding of the deposit. This is in complete agreement with the observed pattern of predominant restriction of microflora to the zone of the oil-water contact, since the surface of contact between oil and water increases during flooding.

As the investigations of Shturm have shown, during the greatest development of the Syzran deposit from 1940 to 1948, the number of bacteria in the formational water increased to six-ten times the previous number, while the flooding of recoverable oil increased to four times the previous value (Shturm, 1950b). We obtained similar data recently on the Pokrovka deposit, where sulfate-reducing bacteria were not generally found by Kuznetsov in 1950 in an oil deposit in the Middle Devonian Bashkirian strata.

At present, because of water injection to maintain pressure, the total number of microflora has increased, and of these, sulfate-reducing microorganisms have become the most abundant. Data no less convincing have been obtained at the Kalinovka deposit of Upper Permian oil, where microflora is more abundant on the western part of the structure, in the zone of intense water exchange, than in other parts of the structure.

The abundance of microflora in oil deposits is variable. For example, bacteria are very abundant in the Lower Carboniferous bed B_2 at the Syzran, Gubino, Radaevka, and Mukhanovo deposits and in the Upper Permian Kalinovka-Novostepanovskoe deposit. Very few bacteria are found in the Kungurian deposits of the trans-Volga region (Vostochnaya Chernovka, Kokhanye, and Yablonevka), in deposits of Bashkirian strata at the Pokrovka deposit, and elsewhere.

It was later ascertained that oil deposits with abundant formational microflora are characterized, as a rule, by artesian water, a fact that indicates good communication between the deposit and formational water. On the other hand, deposits with few formational microflora are either very weakly connected with formational water or are completely barren of formational water; if formational pressure is not maintained by artificial means in this latter situation, the deposit may be developed only with dissolved gas.

An analysis of the geological environment of the investigated deposits has revealed very curious conditions. As has been noted, the oil deposits with intense established water exchange are generally restricted to the sandy unit of the Lower Carboniferous bed B_2, which is regionally developed over a great part of the Ural-Volga region. In contrast to bed B_2, the regional Upper Devonian clastic oil-bearing sequence is hydrogeologically stagnant, according to hydrogeologic studies. There is no doubt that this phenomenon is primarily due to the abundance of microflora in Lower Carboniferous deposits and to the scarcity of microflora in Devonian deposits. In addition, as many investigators have noted, Devonian microflora are not very active, which may be an indication of unfavorable environment.

Previously some microbiologists were of the opinion that formational microflora were absent in Devonian deposits. This barrenness has been explained by the unfavorable physicochemical conditions of the environment, primarily the chemistry of the water. Without denying the abundant grounds for these conclusions, it should be pointed out that the cause of the paucity of bacteria in Devonian strata is associated primarily with very weak water exchange. For example, at the Kuibyshev deposit Ékzertsev and Kuznetsov (1954), Kolesnik (1955), and others have discovered microflora in Devonian deposits of Yablonovka Ravine, at Zol'noe, and at Radaevka. As is well known, the Devonian reservoir rocks of the indicated deposits are highly permeable (up to several hundred millidarcys). Furthermore, faults and abundant fractures have been noted in the indicated areas; these make possible water exchange through discharge of ground water.

The investigations of Meshkov (1958) have shown that, when water exchange occurs, conditions favorable for the development of Devonian microflora may arise. In studying the number of bacteria in the Devonian deposit at Sultangulovo, by direct count, this author computed as many as 273-558 thousand per milliliter in the oil of this deposit. Later, faults were discovered at Sultangulovo, and a hydrochemical survey revealed a hydrochemical anomaly.

Our investigations have shown that there is water exchange through the Devonian strata at Sultangulovo, with the discharge of older water from the Bavly strata, the characteristics of which were established by tests on well 102. It is characteristic that bacteria in the Devonian rocks at Sultangulovo are not fewer than in Carboniferous strata, they are much more abundant than in the Carboniferous and Permian strata of the neighboring Krasnyi Yar deposit, where the number of bacteria is 39-172 thousand per milliliter; only in a single example, in the Lower Permian, does the number reach 316 thousand per milliliter. Recently, microflora have been found in the Devonian deposits of Mukhanovo, Dmitrievka, Chubovka, and other localities; this confirms the wide distribution of microflora in Devonian rocks and the fact of water exchange, although this latter may be very difficult.

No less characteristic is the fact that deposits in carbonate reservoirs generally contain but few bacteria, and in a number of these rocks no primary microflora at all were detected. On the basis of detailed study of the geological structure of the indicated deposits, it has been ascertained that the deposits have been sealed from the underlying formational water by the precipitation of secondary calcite and by viscous bitumens. In developing the Kungurian oil from the Mukhanovo deposit, which is found in dolomite, the formational pressure decreased from 42 atm at the beginning to 4.5 atm. For the period of conservation of the deposit, from November 1955 to the present time, the pressure has remained practically unchanged; this indicates a high degree of isolation from formational water. At the Pokrovka deposit, during the development of the Middle Carboniferous bed A_4, which is limestone, the pressure diminished from 117 atm at the beginning to a bottom pressure (bottom of hole) of less than 56 atm at a number of places. However, the water below did not break through, although it was abundant a few meters below the floor of the deposit.

At the present time it has been reliably established that practically all oil deposits in carbonate reservoir rocks are sealed off on the floor, and any connection between the deposit and circulating formational water is through fractures and faults of later origin.

The scarcity of microflora in oil deposits in carbonate rocks is thus due to this sealing off and the absence of water exchange necessary for the existence of bacteria.

When water exchange is induced artificially in sealed deposits, by pumping water into the deposits, the number of microflora increases markedly. For example, as already noted, the number of bacteria in the sealed deposit in bed A_4 at Pokrovka increased sharply when water was pumped in; during this operation sulfate-reducing bacteria grew very extensively in the flooded zone of the bed (Ashirov, 1959), though they were not previously present in the deposit. The effect of water exchange on kind and number of microflora is equally great in deposits in sandy reservoirs. This is shown by the investigations of Glumov and Stankevich (1959) on the Romashkino deposit in Tataria.

After the first intense pumping of sub-stream ground water directly into the Devonian oil deposit, desulfurizing bacteria began to grow abundantly in the deposit. As a result, hydrogen-sulfide appeared in the Devonian oil deposit.

As is well known, gases dissolved in the Devonian oils of the Ural-Volga region are characterized by the absence of hydrogen sulfide. Its appearance, therefore, in the Devonian deposit at Romashkino must be explained by the artificial effect produced by water exchange as a result of pumping fresh water into the deposit.

In returning to the question concerning the conditions in sealed deposits, it should be noted that the destruction of sulfates in formational water by sulfur-reducing bacteria, with the formation of calcite, has long been known (Ashirov, 1959). However, to evaluate the rate of precipitation of secondary calcite and the possible rate of sealing a deposit, the author, with the microbiologist I. V. Sazonova and the chemist V. P. Parshina, conducted some experiments to study the possibility of precipitating secondary calcite through the influence of formational microflora. The experiments were set up under various conditions at the Kalinovka and Pokrovka oil deposits, with a composite culture of sulfate-reducing bacteria from the Kalinovka formational water.

Abundant precipitation of calcitic sediment occurred during the 80-day period the experiments were continued. According to the petrographers E. K. Frolova and M. M. Larina, the calcite that formed consisted of prismatic crystals 0.01 mm long and less. No calcite formed in control experiments with no sulfate-reducing bacteria.

From these experiments it follows that bacterial precipitation of calcite may be comparatively rapid.

This circumstance may explain why oils in sealed deposits are mostly rather light and geochemically little deteriorated. The good state of preservation of the oil may be explained by rapid closing of any connection between the deposit and formational waters, which, in combination with the microorganisms living in them, are factors favoring active biochemical destruction.

There is interest, both scientific and practical, in investigating the relationship between content of dissolved biogenic gasses (H_2S and CO_2) in oil and formational water and the conditions for bacterial life in a deposit. There is no doubt that formational bacteria are activated when conditions are favorable, and, as a result, biogenic gases

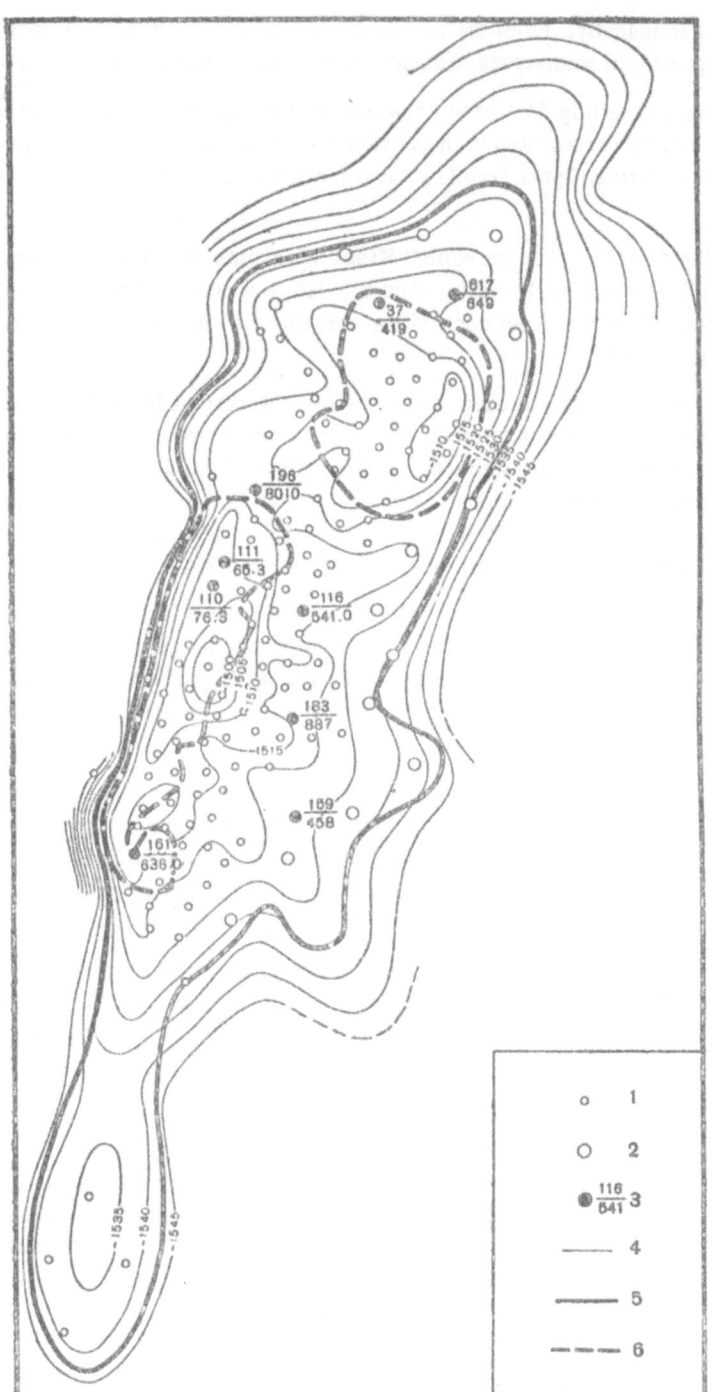

Fig. 1. The Pokrovka deposit. The content of hydrogen sulfide in wells from a coal-bearing horizon. (July-November 1957; scale 1 : 25,000): 1) Producing wells, 2) injection wells, 3) wells in which hydrogen-sulfide content was determined (number of well is the numerator, content of hydrogen sulfide in mg/liter is the denominator), 4) structure contours on the top of bed B_2, 5) outline of the oil accumulation on the roof of the bed before beginning of production; 6) outline of oil zone at the time of investigation.

should be formed more abundantly. From the fact that these gases are present one may infer that the biochemical conditions are favorable and, of primary significance, that water exchange has been intense.

In this connection, interesting data have been obtained in studying the effect of water flooding on the change in degree of hydrogen-sulfide contamination in the flooded section of the Novostepanovskoe deposit. At this deposit, where oil is extracted from Upper Permian Kazanian strata, areal flooding has been carried on in the central zone since 1947.

Extensive pumping of water from the Bolshoi Kinel' River into the indicated strata created conditions favorable for the growth of sulfate-reducing bacteria. And this led to a considerable increase in the hydrogen-sulfide content in the oil and water. Over a ten-year period the content of hydrogen sulfide in the by-product gas increased from 2.2-3.5 to 5-10% and more.

An increase in hydrogen-sulfide content has also been observed at the Pokrovka deposit, where, as a result of pumping water into beds A_4 and B_2, sulfate-reducing bacteria were strongly activated and the content of hydrogen sulfide increased in the by-product gas. Repeated investigations of the gases in bed A_4, taken from deep oil samples, have shown that the content of hydrogen sulfide content at well 1 was 0.7% (by volume) on December 3, 1950 and 0.9% on November 1, 1957. In well 10 the content was 0.9 on May 27, 1950, 1.1% on November 17, 1956. In well 11 the value was 0.2% on December 7, 1950, 0.4% on October 23, 1957, and so forth.

A clear relationship between hydrogen-sulfide content in extracted liquid and position of wells in the zone of flooding may also be seen in bed B_2 (Fig. 1). In wells in the flooded zone the content of hydrogen sulfide, according to our data, is 410-887 mg/liter, but in the zone of pure oil only 65-92 mg/liter. In this regard it should be noted that at the time of the investigation the front of fresh water, pumped from the perimeter, had not yet reached any of the flooded wells.

The short-period pumping of water with 80-900 mg and more of formaldehyde per liter, which we directed at the Kalinovka-Novostepanovskoe deposit, confirmed the fact that it is possible to destroy sulfur-reducing bacteria in a bed completely and to decrease somewhat the content of hydrogen sulfide (Ashirov, Gromovich, Parshina, Sozonova, 1959). However, the practical use of the indicated treatment depends on the search for a more effective and inexpensive antiseptic.

Experience at the Romashkino and other deposits shows that antiseptic treatment of the water pumped directly into the deposit is required at the beginning of flooding. Only in this way will a safe front be created at the contact between oil and water, where microflora generally develop actively; it is the effect of this activity wish to be delivered from.

Confirmation of increased generation of carbon dioxide in the oil-water contact zone has been found in the work of Kozin and Mzhachikh (1958) on the composition and content of gases dissolved in waters of oil deposits in the Kuibyshev district. For example, in formational water next the oil in the Devonian rocks of the Zol'noe deposit the gas amounts to 0.469 m^3/m, and CO_2 constitutes 6.1-13.1% (by volume); in the zone beyond the contact the gas factor declines to 0.298 m^3/m, and the CO_2 drops to 4.9%. In the Lower Carboniferous bed B_2, the gas is 1.683-1.738 m^3/m, and the CO_2 content is 7.5-11.2%; and beyond the contact this factor is but 0.248-0.392 m^3/m, and the CO_2 content is only 3.36-4.20%.

Interesting data have been obtained for the zone outside the contact in hole 21 at Pokrovka. The hole is isolated from the oil deposit in bed B_2 by a tectonic barrier. It is obvious, therefore, that CO_2 is absent in the dissolved gas, at a gas factor in the water of 0.202-0.283 m^3/m.

The cited examples confirm the view that the generation of CO_2 is confined to the zone of the oil-water contact, i.e., to the zone of active biochemical processes.

Thus, the content of biogenic gases in formational oil and water is associated to a considerable extent with biochemical activity, which is determined by the intensity of the water exchange, i.e., the rate of ground-water flow.

To investigate the possible biochemical conditions of the medium, we compared two deposits in the Kuibyshev segment of the trans-Volga region: the Krasnyi Yar and the Belozerka (Fig. 2).

Both deposits began development in 1958-59; they produce from the same sandy bed (B_2), and the quality of oil and composition of ground water are practically identical. However, the gas factors and the content of CO_2 and

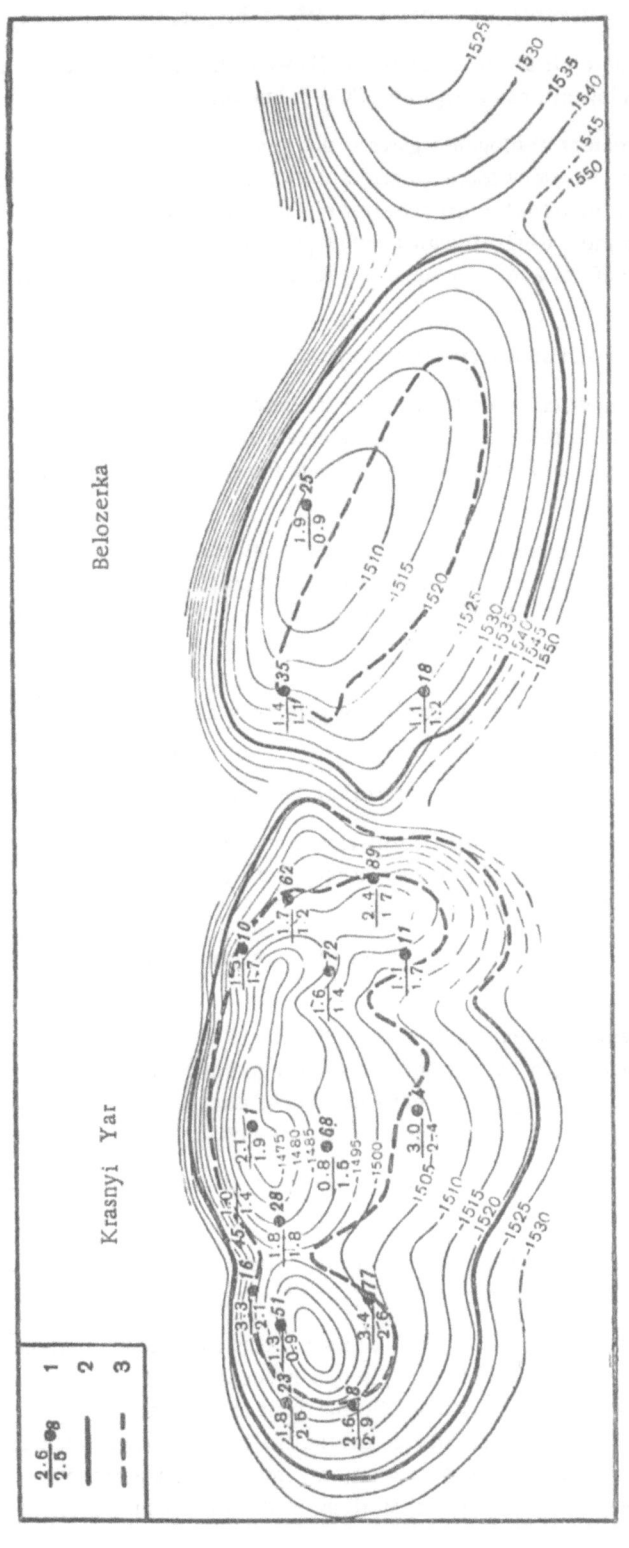

Fig. 2. Content of H_2S and CO_2 in the gases of bed B_2 at the Krasnyi Yar and Belozerka deposits: 1) Content of hydrogen sulfide (numerator) and carbon dioxide (denominator) in molar percent; 2) outline of oil zone, drawn on top of bed; 3) outline of oil zone, drawn on bottom of bed.

H_2S in the gases differ sharply. The gas factor of the Krasnyi Yar deposit is 49 m^3/ ton, that of Belozerka 29 m^3/ton. At Krasnyi Yar the H_2S content in gases from wells in the oil zone ranges from 0.8 to 1.8%, and in wells of the water-oil zone, from 1.8 to 3.3%. The content of carbon dioxide in gases from the oil zone ranges from 0.8 to 1.8%, and from the water-oil zone, 2.1-2.9%. At Belozerka H_2S forms 1.1-1.9% of the gas, CO_2 0.9-1.2%.

As may be seen, at Belozerka, where all the investigated wells were in the flooded zone, in addition to a much lower gas factor, the content of biogenic gases (H_2S and CO_2) is but one-half to one-third that at Krasnyi Yar.

The cause of this difference in content of biogenic gases is not yet known. However, the available data permit us to conclude tentatively that there has occurred a deterioration of the hydraulic connection between the Belozerka deposit and the surrounding formational artesian-water system (as compared with the Krasnyi Yar deposit). In respect to this, it may be said that a more difficult water exchange in the Belozerka area created poorer conditions for the growth of formational microflora and, consequently, for the accumulation of biogenic gases.

In this conclusion concerning the Belozerka deposit is confirmed, we may then form judgments on the openness of the connection between a deposit and an artesian-water system by determining the content of biogenic gases.

The possibility of answering this question beforehand, even if approximately, at the conclusion of deep exploration will aid during planning operations to foresee the rate of decrease in formational pressure for a given level of petroleum production and to provide at an early stage for maintaining the formational pressure or to reject the operation entirely.

If we consider that the cost of flooding outside the oil boundary may reach 30% of the total expenditure for construction, and that for a deposit of average size this may amount to tens of millions of rubles, it then becomes clear how necessary it is to find an immediate solution to this problem.

Conclusions

1. The degree of development of formational microflora is primarily associated with the intensity of formational water exchange.

When formational water is stagnant, the number of microflora becomes much smaller.

2. When water exchange is stimulated artificially, formational microflora become more active and more numerous. This may occur not only during flooding of a bed by surface water, but also when the water exchange is stimulated by introducing highly mineralized water.

3. The abundance and activity of microflora in an oil deposit may serve as an index for the status of deep water exchange.

4. It has been ascertained that secondary calcite will precipitate and tarry oils (up to viscous bitumens) settle out, through the action of formational microflora, in the oil-water contact zone of deposits in carbonate reservoir rocks. In this connection, completely formed deposits in carbonate rocks are generally sealed on the floor from formational water. This phenomenon is also observed, though much less thoroughly developed, in sandy reservoir rocks.

5. The scarcity of formational microflora in sealed deposits may be an index to the degree of isolation of the deposits from formational water.

6. The degree of isolation of a deposit from an artesian-water system may be determined by the content of biogenic gases (CO_2 and H_2S) in the by-product gases.

7. An increase in hydrogen-sulfide contamination of a deposit during injection of surface water, especially when there is areal flooding and flooding within the oil zone, makes it necessary to treat the water with antiseptics that can prevent the growth of microflora in the formations. Primarily it is necessary to suppress the growth of sulfate-reducing bacteria, since the increase of hydrogen-sulfide contamination of the oil produces marked corrosion of the equipment, contaminates the atmosphere about the operations, impairs the treatment of the oil, and makes it necessary to conduct expensive supplemental purifying operations on the by-product gases before supplying them to consumers.

8. In connection with the established fact that bacteria seal oil deposits in carbonate reservoir rocks, and for the purpose of gaining the maximum oil production under these circumstances, it is necessary to maintain the pressure in these rocks by injecting water during the initial stages of development of the deposit. When oil deposits are sealed off, flooding from beyond the oil perimeter is impossible.

The water should be pumped directly into the deposit in the oil-water contact zone, or over the entire area by means of a series of injection wells. In places the above-indicated conclusions also apply to deposits in sandy reservoir rocks.

LITERATURE CITED

Ashirov, K. B. 1959. "Cementation of the bed near the contact of oil deposits in carbonate rocks and the effect of this process on development of the deposits. Geology and development of oil deposits." Trudy Inst. Giprovostokneft', No. 2.

Ashirov, K. B., Gromovich, V. A., Parshina, V. P., and Sazonova, I. V. 1959. "Results of experiments on diminishing the content of hydrogen sulfide in the recovered fluid at the Kalinovko-Novostepanovskoe deposit. Geology and development of oil deposits." Trudy Inst. Giprovostokneft', No. 2.

Ekzertsev, V. A. 1951. "Microscopic studies of bacterial flora in the oil-bearing facies of Vtoroi Baku." Mikrobiologiya, 20, No. 4.

Ekzertsev, V. A. and Kuznetsov, S. I. 1954. "Studies of the microflora in the oil deposits of Vtoroi Baku." Mikrobiologiya, 23, No. 2.

Glumov, I. F. and Stankevich, E. F. 1959. "The appearance of hydrogen sulfide in injection wells of the Romashkino deposit." Tatarskaya neft', 1-2.

Kolesnik, Z. A. 1955. "Microflora in water and oil of parts of Vtoroi Baku. Conditions for the formation of oil." Trudy VNIGRI, novaya seriya, No. 82.

Kozin, A. N. and Mzhachikh, K. I. 1958. "The composition of gases in formational waters of oil deposits in the Kuibyshev district. Geology and development of oil deposits." Trudy Inst. Giprovostokneft', No. 1.

Kuznetsov, S. I. 1957. "Principal results of studying the microflora of oil deposits." Mikrobiologiya, 26, No. 6.

Kuznetsova, V. A., Ashirov, K. B., Gromovich, V. A., Ovchinnikova, I. V., and Kuznetsov, S. I. 1957. "An experiment in suppressing the growth of sulfate-reducing bacteria in the oil bed at the Kalinovka deposit." Mikrobiologiya, 26, No. 3.

Meshkov, A. N. 1958. "Using the direct count for studying microflora in oil." Mikrobiologiya, 27, No. 3.

Shturm, L. D. 1941. Microorganisms in the Oil Deposits of Vtoroi Baku [in Russian], abstract of work at the Institute of the Division of Biological Sciences, Academy of Sciences, USSR, for 1940, Izd. AN SSSR.

Shturm, L. D. 1950a. "Microscopic studies of oil-bearing beds and water." Mikrobiologiya, 19, No. 1.

Shturm, L. D. 1950b. "Data on the microbiological studies of oil deposits in Vtoroi Baku." Trudy Inst. nefti, No. 3.

RESULTS OF STUDIES ON FORMATIONAL MICROFLORA
IN THE OIL DEPOSITS OF THE KUIBYSHEV REGION

I. V. Sazonova

(Scientific-Research Institute of the Petroleum Industry, Kuibyshev)

A study of microfloral groups in an oil bed has been made at 18 deposits in the Kuibyshev region. Special attention was directed toward the microflora in young freshly drilled deposits. The microflora from deposits that have produced for a long time was also studied. Samples of formational water and oil were analyzed for sulfate-reducing, butyric-acid, methane-forming, sulfur-oxidizing, and denitrifying bacteria and for microorganisms that oxidize gaseous hydrocarbons and hydrogen and that destroy oil in an anaerobic environment. Nutrient media were prepared for each culture according to the mineralization and type of salt in the specific deposit and bed.

Parallel inoculations were made on ordinary culture media with the addition of 1 to 15% NaCl.

The investigated groups of microorganisms grew best, as a rule, on media containing from 1 to 1.5% NaCl. Rarely were strains distinguished that grew at the mineralization and salinity of the formational waters.

The data we obtained confirm those of Shturm (1950) and Ékzertsev and Kuznetsov (1954) for these areas. Sulfate-reducing and methane-forming bacteria represent the groups most frequently encountered in the oil bed. Denitrifying bacteria were also found in considerable numbers.

The presence of sulfate-reducing bacteria is in agreement with the presence of hydrogen sulfide in the formational waters. This is clearly seen at the new deposits (Krasnyi Yar, Mikhanovo). Young, freshly drilled deposits are characterized by a scarcity of microorganisms: few groups and few individuals.

The greatest number of microorganisms is found in old flooded deposits; at such cites one may frequently observe butyric-acid bacteria and the group Pseudomonas, which leads to the anaerobic destruction of oil.

However, butyric bacteria are sometimes encountered in new deposits as well.

In studying the microflora of the Mukhanovo deposit, we separated cultures that grew on oil as the only source of carbon.

Denitrifying bacteria, which are frequently observed in the formational waters of Devonian deposits, are widespread at oil fields.

It is very important to analyze samples of formational water and oil immediately after they are collected or, at least, within two weeks. The storage of samples leads to changes in the natural relations between groups of microorganisms.

In studying the microflora from formational waters in wells with artesian head, it is necessary to collect samples merely from the depth of perforation. Samples from other depths generally contain those groups that enter the hole from other formations.

For example, wells in Devonian rocks with artesian water at the Mukhanovo deposit contain sulfate-reducing bacteria, although this group is found in bed I of the Lower Carboniferous.

Before collecting microbiological samples from wells with artesian water, the wells must first be flushed.

Microflora are good indicators of the chemical composition of formational waters. Sulfate-reducing bacteria have a characteristic distribution in the formational waters of the Pokrovka deposit.

At this deposit the bacteria in individual wells correspond closely to the composition of the formational waters. Wells in Bashkirian strata (Nos. 226, 563, 210, and others), in which the mineralization is low (43.7, 58.5 mg-equiv per 100 cm^3), with a high content of sulfates and a low content of bromine and iodine, have a maximum number of sulfate-reducing bacteria.

On the other hand, in wells with high mineralization but low sulfate content and high bromine and iodine content, sulfate-reducing bacteria are absent.

This same relationship is also noted for a coal-bearing horizon. The growth of sulfate-reducing bacteria in certain wells marks the approach of injected fresh water to these wells.

At cites of the oil-water contact in oil deposits, one may frequently discover deposits of secondary calcite. The place of its deposition attests to a biogenic origin. We have attempted to reproduce in the laboratory the process of sealing a deposit by means of sulfate-reducing bacteria. Sulfate-reducing bacteria are generally found in oil deposits, and most frequently at the oil-water contact.

Experiments were set up with Tauson medium and various sources of hydrogen, such as calcium lactate, calcium acetate, oil from different deposits, and gaseous hydrogen.

The experiments were performed with sterile tubes, stoppered at both ends by stoppers with gas-vent tubing, and having a porous glass partition on which was placed a piece of core saturated with oil but containing no traces of calcite. The tube was filled with some nutrient solution.

A composite culture of sulfate-reducing bacteria was taken from the same deposit as the oil. The oxidation-reduction conditions and the acidity in the experiment corresponded to the actual environment in the sedimentary formation.

The experiments showed that sulfate-reducing bacteria, in their life processes, lead to the formation of secondary calcite similar to the secondary calcite found in oil deposits.

Conclusions

1. A study has been made of the microflora from formational waters of 18 deposits in the middle Volga region.

2. The greatest number of microflora were found in deposits that had been worked for many years.

3. Sulfate-reducing and methane-forming bacteria and bacteria leading to the anaerobic decomposition of oil were encountered more frequently than other groups.

4. Hydrogen sulfide in formational waters is associated with sulfate-reducing bacteria.

5. Laboratory experiments have shown that secondary calcite, which is frequently found at the oil-water contact and which seals oil deposits, may be of biogenic origin.

LITERATURE CITED

Ékzertsev, V. A. and Kuznetsov, S. I. 1954. "An investigation of the microflora in the oil deposits of Vtoroi Baku." Mikrobiologiya, 23, No. 1.
Shturm, L. D. 1950. "Data on the microbiological investigation of oil deposits in Vtoroi Baku." Trudy Inst. nefti. 1, No. 1.

DISTRIBUTION AND ECOLOGY OF MICROORGANISMS
IN THE DEEP GROUND WATERS OF SOME REGIONS OF THE USSR

Z. I. Kuznetsova

(All-Union Scientific-Research Institute of Hydrogeology
and Engineering Geology, VSEGINGEO, Moscow)

There is practically no information, either in domestic or in foreign literature, on the role of bacteria in ground water in changing the chemical composition of the water down the dip of aquifers. The first steps in this direction were taken at the All-Union Scientific-Research Institute of Hydrogeology and Engineering Geology (Al'tovskii, Shvets, and Kuznetsova, 1958; Kuznetsova, 1960).

It should be noted that, in setting up an operation to investigate this matter, great difficulties arise in collecting samples. Samples for bacterial analysis were collected from available springs in the region of investigations or from oil and gas wells. In a number of places, where it has been impossible to trace the changes in bacterial content of the water along a single aquifer from recharge zone to discharge zone, conclusions are drawn from average data for samples collected in the recharge zone of aquifers, in zones of stagnant water at oil deposits, and in discharge zones.

The bacterial analysis of the water was made by a quantitative count directly under the microscope and by segregation of composite cultures of the most interesting physiological groups.

The intensity of bacterial growth according to physiological groups has been indicated tentatively on a five-division scale. The basis of evaluation is the time of appearance of marked signs of growth, detected by the unaided eye, in the composite culture. A scale value of 5 is given to samples in which growth begins within five days, a value of 1 to samples in which marked development of bacteria begins in the period of 40 to 60 days.

In this technique, active cultures of bacteria appear quickly and are given high scale values. In studying the microflora of deep waters, active cultures of bacteria are especially important, since, because of the specific conditions in the surrounding medium of the ground water, one might expect an accumulation of bacterial cells having no effect on the change of chemical composition of the water.

In computing the total number of bacteria, direct count under the microscope was made. However, as is well known, the mineralization of ground water increases notably with depth of aquifer, amounting to as much as 200-300 g/liter in some places, and the temperature of the water may reach 100° and more. In samples from such waters one may list, in direct count, dead cells along with viable cells.

Therefore, in making quantitative computations of bacteria from ground water, separate counts were made for live and dead bacteria by staining (according to Peshkov) with Giemsa's solution after staining with lichtgrun according to the method developed by Lazarev (1953).

The objective of the present investigations has been the solution of a number of questions concerning material obtained from ground water in the Grozny-Dagestan province:

1. To discover if there is a change in the proportions of viable bacterial cells along an aquifer;

2. to show how the bacterial population of the water changes along the water-bearing horizon of the Chokrak bed C at the oil fields of Ternair and Makhachkala;

3. to gain an idea of the changes in bacterial population of the water from recharge zone to discharge zone on the basis of average data from a large number of analyses.

From the accompanying Table 1 it may be seen that the total number of bacteria ranges from 118,000 to 634,000 per milliliter of water in samples collected in the foothills of Dagestan, in the zone where the rocks are exposed. Almost all the bacteria detected in these waters were stained by Giemsa's solution as live cells, the average percent of viable cells reaching 99%.

In samples collected at the oil field in Makhachkala, the total number of bacteria ranged from 10,000 to 620,000 per milliliter of water.

Two groups of samples may be noted here, according to the number of viable bacterial cells: samples collected from wells with large yield (45 tons per day) and samples from low-yield wells (from 2 to 7 tons per day) containing about 50% oil.

Most of the samples belong to the first group, in which the percent of viable cells reaches 94. In the second group the percent of viable cells is 7, on an average.

Apparently physicochemical conditions develop in the shaft of a low-yield well that differ sharply from the natural conditions in formational water, possibly because of an increased proportion of oil, which causes the bacteria to die rapidly. At gas deposits the number of viable cells, regardless of the yield of the well, was found to be high.

TABLE 1. Change in Content of Viable Bacterial Cells (ground water of Dagestan ASSR)

Sample locality	$t°C$	Yield*	Total No. of bacteria per ml, in thous.	Average% of living cells	No. of analyses	Depth, m
Recharge zone (springs)	11.5-19.5	0.5-0.2	118-634	99	3	
Makhachkala oil field	42-62	45-17,100	10-620	94	6	1200-1900
(wells)	32	2-7.5	193-212	7	2	1400-1960
Gas deposit (wells)	29-41	17-1.1	10-242	98	4	225-225 †
Discharge zone (springs)	14-86	0.1-6	36-1900	97	5	

* Yield for springs in liters/sec, for wells in m^3/day.

On the average 98% of the cells were stained by Giemsa's solution as living. Apparently an increase in content of gas in a sample does not affect the viability of the bacteria.

The following series of samples come from mineral springs in the discharge zone of aquifers.

As may be seen in Table 1, the total number of bacteria in these samples ranged from 36,000 to 1,900,000 per milliliter of water, but the percent of viable cells was, on the average, 97.

On the basis of the work done, it may be stated that, as a rule, bacterial populations in water in regions of intense water exchange, regardless of depth of aquifer, are active; only rarely, when the natural conditions of the environment are destroyed, to the bacteria die out.

Let us turn to an examination of the results obtained during analysis of bacterial populations of water down the dip of an aquifer in the oil fields of Ternair and Makhachkala.

Figure 1 is a diagram showing the changes in total number of bacteria and in some of the physiological groups in relation to the ecological conditions.

As the diagram shows, the temperature of the water increases from 18 to 48° with depth of burial of the aquifer, from the recharge zone (a spring on the road from Makhachkala to Talgi; sample 1) to a depth of 1600 m (Ternair; well 10); between these two localities the total number of bacteria drops from 660,000 to 42,000 per milliliter of water.

† As in original.

In the next three wells the temperature of the water ranged from 50 to 60° and the total number of bacteria ranged from 32,000 to 70,000 per milliliter of water. And only in a sample collected at the Makhachkala field, from well 76, where the temperature of the water was not high (25°), did the number of bacteria become as great as 840,000 per milliliter of water. There is thus observed an inverse relationship between number of bacteria and temperature down the dip of the aquifer.

Let us turn to an examination of the data on intensity of growth of desulfurizing bacteria.

Sulfates, organic material or molecular hydrogen, and a low oxidation-reduction potential are necessary for the active growth of desulfurizing bacteria.

These bacteria develop very weakly in the recharge zone of aquifers where $rH_2 = 24$. Their intensity of growth here is assigned a value of one on the scale. Their abundance in the zone of oil deposits, in samples collected from well 10 at the Ternair field and from well 76 at the Makhachkala field, reaches 4 and 5 on the scale.

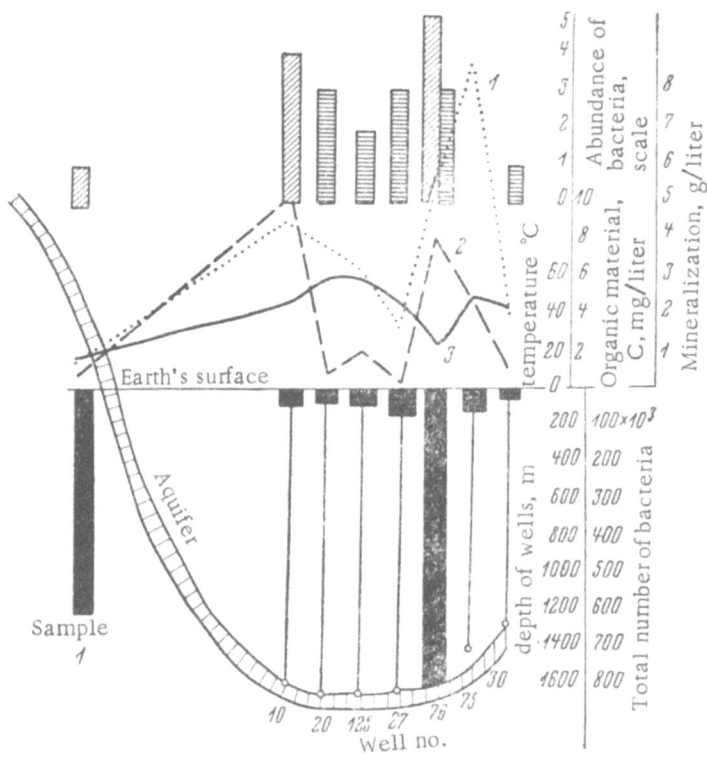

Fig. 1. Diagram showing bacterial population in water down the dip of an aquifer in Chokrak strata at the Ternair and Makhachkala fields: 1) Mineralization, g/liter; 2) organic carbon C, mg/liter; 3) temperature, °C; columns with oblique ruling; sulfur-oxidizing bacteria; solid black columns; total number of bacteria.

As seen in Fig. 1, the abundance of desulfurizing bacteria is found in samples with a high content of organic material. In samples collected from wells 10 and 76, the organic content reached 8 and 10 mg/liter, whereas in the other samples it was near 2 mg/liter.

In studying the abundance of sulfur-oxidizing bacteria, determinations of rH_2 were made.

Unfortunately, in this series of samples the oxidation-reduction potential of the environment was determined only in sample 1 and in well 30, where it ranged from $rH_2 = 24$ to $rH_2 = 17.8$.

However, data obtained from a number of samples in the Grozny-Dagestan oil province make it possible for us to stage that the abundance of sulfur-oxidizing bacteria belonging to the species <u>Thiobacillus</u> <u>thioparus</u> is related definitely to the oxidation-reduction potential of the environment.

The abundance of sulfur-oxidizing bacteria indicated by one on the scale was observed in the recharge zone of aquifers in but a single sample, where $rH_2 = 21$; in the other 7 samples, where rH_2 ranged from 23 to 30, the growth of sulfur-oxidizing bacteria was not observed (Fig. 2).

At oil deposits where the oxidation-reduction potential of the medium ranged from $rH_2 = 13$ to $rH_2 = 18$, the abundance of sulfur-oxidizing bacteria measured 3-5 on the scale. When the value dropped to 9.7-10.6 at these sites, the bacteria did not grow. Consequently, sulfur-oxidizing bacteria grew actively at values of rH_2 ranging from 12 to 18.

In springs at the discharge zone, as well as in oil-bearing waters, the lower boundary of growth of sulfur-oxidizing bacteria lies at $rH_2 = 11.6$-12.6.

On the basis of work done, we may conclude that the abundance of sulfur-oxidizing bacteria down the dip of an aquifer is dependent on the oxidation-reduction potential of the environment.

On the basis of analyses of a great number of samples, it may be seen that changes in bacterial content of water in relation to the ecology may be followed as well from analyzing average values as by analyzing samples collected from a single bed.

Average data on changes in chemical composition and bacterial content of water in the recharge zone of aquifers, in the ground water at oil deposits, and in the discharge zone of ground water are given in Table 2; from these it may be seen that the temperature of the water is characteristically low in the recharge zone, the water commonly contains soluble hydrogen, and the rH_2 ranges from 17.8 to 30. The total number of bacteria in the water ranges from 59,000 to 1,000,000, depending in great measure on the yield of the spring and on whether the spring is plugged (or dammed).

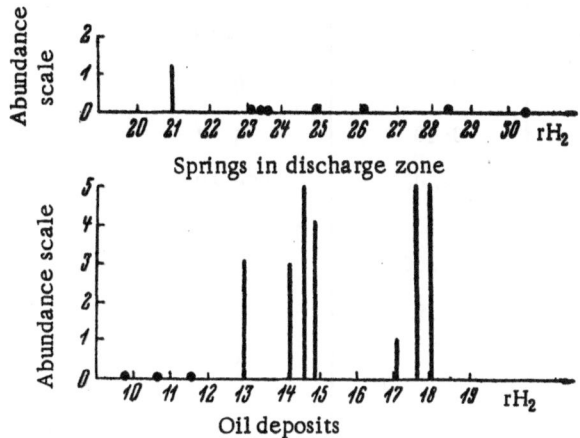

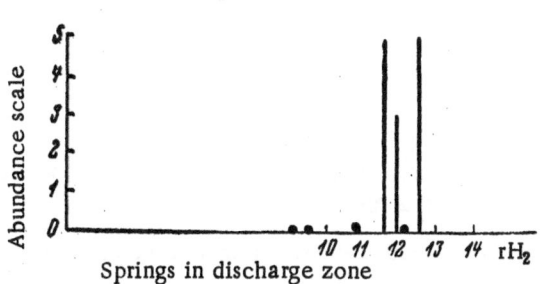

Fig. 2. Abundance of sulfur-oxidizing bacteria in relation to the rH_2 of the environment.

Desulfurizing and sulfur-oxidizing bacteria are weakly developed, a fact apparently associated with the high oxidation-reduction potential of the medium. Putrefying bacteria that destroy albuminous material grow well.

With increasing depth of the aquifer, the temperature of the water increases and reaches values of 89° at places in the zone of oil deposits. The oxidation-reduction potential of the environment drops, varying between the limits of rH_2 of 9.7 and 17.8, soluble hydrogen disappears, hydrogen sulfide appears in the water in amounts up to 40 mg/liter, and the over-all mineralization of the water increases. In these waters one observes an inverse relationship between total content of bacteria and temperature of the water. At temperatures ranging from 12 to 21°, the total number of bacteria ranges from 105,000 to 9,000,000. At temperatures between 50 and 89° the number of bacteria drops to 25-72,000 per milliliter of water. With a drop in the oxidation-reduction potential in the water, desulfurizing and sulfur-oxidizing bacteria begin to grow. The abundance of putrefying bacteria declines.

And, lastly, samples of water were collected in the discharge zone from springs, where the temperature of the water ranged from 12 to 87°. The oxidation-reduction potential ranged from RH_2 of 7.6 to 15; dissolved hydrogen sulfide was detected in all samples. An inverse relationship between total content of bacteria in the water and temperature of the water was also observed in the discharge zone. At temperatures up to 50°, the content of bacteria ranged from 114,000 to 3,000,000. At higher temperatures, between 50 and 89°, the number of bacteria in the water dropped to 9-40,000 per milliliter of water. The abundance of sulfate-reducing and sulfur-oxidizing bacteria increased to 3-4 on the scale. The number of putrefying bacteria was somewhat less than that for oil-bearing waters.

TABLE 2. Average Values of Changes in Chemical Composition and Bacterial Content in Aquifers

Indicator	Sample locality		
	recharge zone	oil deposits	discharge zone
Temperature, °C	12-21	23-89	12-87
rH_2	30-17.8	17.8-9.7	15-7.6
O_2, mg/liter	5.6-0.6	0	0
H_2S, mg/liter	0-0.2	0-40	4-57
Total mineralization, mg/liter	60-1570	1000-5000	820-6240
Number of bacteria per ml at t < 50° (in thousands)	59-1036	105-9600	144-3137
The same at t > 50°	—	25-72	9-40
Desulfurizing *	0-1	2	3
Sulfur-oxidizing	0-1	2	4
Putrefying	3	0-1	1

* Abundance of bacteria indicated on conventional scale.

Conclusions

1. The bacterial population of ground water is generally active, and only under certain circumstances, when the natural conditions of the environment are sharply disturbed, do the bacteria die out.

2. The total number of bacteria down the dip of an aquifer decreases with mounting temperature of the water.

3. The abundance of desulfurizing bacteria at oil deposits corresponds to an increased content of organic material in the water.

4. As a rule sulfur-oxidizing bacteria are found to be very abundant where the oxidation-reduction potential of the water is between rH_2 value of 11.6 and 21.

LITERATURE CITED

Al'tovskii, M. E., Shvets, V. M., and Kuznetsova, Z. I. 1956. Origin of Oil and Oil Deposits [in Russian] (English translation available from Consultants Bureau Enterprises, Inc.).

Kuznetsova, Z. I. 1960. "A study of the distribution of desulfurizing bacteria down the dip of an aquifer in ground water of the Terek-Degestan oil province." Vopr. Gidrogeologii i inzhenernoi geologii, Collection 18, VSEGINGEO.

Lazareva, M. F. 1953. Direct Bacterial Count for Solving Problems in the Technical Microbiology of Water [in Russian], Izd. VODGEO.

DISTRIBUTION AND SPECIFIC CONTENT OF BACTERIA THAT OXIDIZE GASEOUS HYDROCARBONS IN THE GROUND WATER AT GAS DEPOSITS IN THE AZOV-KUBAN BASIN

Z. P. Telegina

(All-Union Scientific-Research Geological-Prospecting Petroleum Institute, Moscow)

Systematic investigation of ground water for the presence of bacteria that grow in an atmosphere of methane, ethane, propane, or butane was first made in 1958-59 under the direction of G. A. Mogilevskii at the gas fields of the Azov-Kuban basin, including the Leningradskaya, Starominskaya, Kanevskaya, and other areas.

The structures of the Azov-Kuban basin are made up of Cretaceous, Tertiary, and Quaternary rocks. Both Tertiary and Quaternary rocks are chiefly sequences of sandy and clayey sediments

The principal object of study was water from wells (pumping wells, artesian wells, and, in part, dug wells) tapping upper Tertiary and Quaternary aquifers. The depth of the investigated aquifers nowhere exceeded 250 m.

The waters of the investigated region are fresh, and are chiefly of two types: sodium bicarbonate and sodium sulfate. The waters of the Pliocene and Quaternary rocks of the Azov-Kuban basin are rich in microflora of extremely varied species. An average of 75,000 to 500,000 cells were counted on membrane filters from 1 ml of water. The method used in analyzing the samples of water for the presence of hydrocarbon-oxidizing bacteria is as follows: 2.5 ml of the investigated water was added to the same quantity of liquid mineral medium, poured into a soil pot with sterile sand on the bottom. A gas mixture, containing one-third nutrient gas and two-thirds air, was supplied as the source of nutrient hydrocarbons.

The bacteria were cultured at a temperature of 32-35° for 14 days.

Data on the distribution of hydrocarbon-oxidizing bacteria in the individual areas of the Azov-Kuban basin are shown in Table 1.

As may be seen from the data in Table 1, methane-oxidizing and propane-oxidizing bacteria proved to be the most widespread groups of microorganisms.

Ethane- and butane-oxidizing bacteria are much more rarely encountered in the waters of the Tertiary and Quaternary rocks of the Azov-Kuban basin.

In examining the distribution of methane-, ethane-, propane-, and butane-oxidizing bacteria along upper Tertiary and Quaternary aquifers, it may be seen that the number of bacteria that oxidize methane, propane, and butane are much more abundant in the waters of upper Tertiary rocks than in waters of Quaternary rocks (Table 2); this relationship is in keeping with the analyses of hydrocarbons.

A study of the combinations found among the above-described groups of hydrocarbon-oxidizing bacteria has shown that bacteria that oxidize individual gaseous hydrocarbons, such as methane, are found in 23% of the water samples from which other hydrocarbon-oxidizing bacteria are absent. The most common combinations (up to 15%) of the bacterial groups in waters of the Azov-Kuban basin were found to be species of methane- and propane-oxidizing bacteria. Other combinations of hydrocarbon-oxidizing bacteria were encountered more rarely.

Sixteen strains of bacteria were separated in pure cultures for the purpose of studying the specific content of the hydrocarbon-oxidizing bacteria.

TABLE 1. Distribution and Abundance of Bacteria that Oxidize Gaseous Hydrocarbons in Certain Areas of the Azov-Kuban Basin (abundance indicated in conventional units)

Area	No. of water samples analyzed	Presence of bacteria that oxidize							
		methane		ethane		propane		butane	
		% of samples in which detected	abun-dance	% of samples in which detected	abun-dance	% of samples in which detected	abun-dance	% of samples in which detected	abun-dance
Leningradskaya	10	60	25.9	20.0	13.5	80	59	10	0.4
Starominskaya	15	66.6	122.6	13.3	4.5	40	39.1	53.3	25.3
Kanevskaya	19	52.6	14.1	0	0	57.9	35.3	15.8	2.5
Novoshcherbinovskaya	7	85.7	31.4	0	0	28.5	4.5	14.3	14.3
Eiskii region	10	30.0	3.4	0	0	20	0.6	0	0
Azov-Kuban basin as a whole	61	58.98	39.5	6.65	3.6	45.3	27.5	18.7	8.5

Of these, three strains were isolated in an atmosphere of methane. One species was obtained from water in Quaternary rocks, and two strains came from core samples of Cretaceous rocks, from depths of 2070 and 2075 m. In morphological, cultural, and physiological features, the first strain belongs to Methanomonas methanica; the other two are assigned to Bacterium methanica.

Eight strains of bacteria were isolated in an atmosphere of propane. All the strains were obtained from aquifers in Tertiary strata except for one culture of Mycobacterium rubrum, which was isolated from water in Quaternary rocks. From their morphological and physiological properties, some of the propane-oxidizing bacteria were referred to the genus Pseudomonas, specifically to Ps. pantotropha and Pseudomonas sp. (not identified more precisely).

TABLE 2. Distribution of Hydrocarbon-Ozidizing Bacteria in Upper Tertiary and Quaternary Aquifers of the Azov-Kuban Basin

Age of aquifer	No. of samples	Ave. content of saturated hydrocarbons in the investi-gated waters, ml/ liter	No. of water samples (%) in which bacteria grew in an atmosphere of			
			methane	ethane	propane	butane
Quaternary	15	0.35	46.6	6.6	40	6.6
Upper Tertiary	45	0.72	62.2	6.6	48.8	26.6

Some of the microorganisms were representatives of the species Mycobacterium rubrum var. propanicum, M. carotenum, and Mycobacterium sp.

And, lastly, one strain was referred to Pseudobacterium subluteum.

Five strains were isolated in an atmosphere of butane; these belonged to four species of bacteria: Pseudobacterium subluteum, Pseudomonas fluorescens, Pseudomonas, sp., and Actinomyces condidus.

On the basis of the data obtained, it is seen that the greatest number of bacteria capable of oxidizing gaseous hydrocarbons belong to the genus Pesudomonas. Hydrocarbon-oxidizing bacteria of the genera Mycobacterium and Pseudobacterium are less abundant.

Conclusions

1. Methane- and propane-oxidizing bacteria are the most widespread groups of microorganisms in waters of the Tertiary and Quaternary rocks in the gas district of the Azov-Kuban basin.

2. Bacteria that oxidize gaseous hydrocarbons are represented by the genera Pseudomonas, Mycobacterium, Pseudobacterium, and Actinomyces.

3. Most of the microorganisms using gaseous hydrocarbons in the investigated region belong to the genus Pseudomonas.

4. A correspondence between bacterial distribution and data from gas analyses, on the one hand, and a great variation in species, on the other, point to active geologic work of these microorganisms, leading to the oxidation of natural gases.

THE ROLE OF BACTERIA IN THE OXIDATION OF SULFIDE ORES

N. N. Lyalikova

(Institute of Microbiology, Academy of Sciences, USSR, Moscow)

An explanation of the role of microorganisms in the alteration and transformation of mineral deposits is one of the principal objectives of geological microbiology. Many valuable metals, such as copper, zinc, nickel, molybdenum, and others occur in nature chiefly in sulfides; it is therefore important that we study the oxidation processes of sulfide ores. This problem has received considerable attention from geologists, an example being the classic monograph of S. S. Smirnov (1955). But all the work on the oxidation of sulfides has considered the matter purely from the chemical point of view. Investigations of the role of bacteria in the oxidation of sulfide ores have been very scanty until recently. The effect of bacteria in the oxidation of pyrite was studied by Rudolfs (1922), who, with Helbronner, also studied the effect of bacteria on the leaching of zinc from zinc blende (Helbronner and Rudolfs, 1922). As a result of the work of Rudolfs and Helbronner, it has been ascertained that the oxidation process is biogenic, but it is not yet clear which bacteria act on pyrite and zinc sulfide or to what degree the bacteria accelerate the oxidation of these minerals.

In studying the biological oxidation of sulfides, an important agent was brought to light by the discovery of a new autotrophic organism Thiobacillus ferrooxidans, which was first isolated by the American investigators Colmer and Hinkle (1947) in the acid waters draining coal mines.

In many coal beds there are inclusions of iron sulfide: pyrite or marcasite. As a result of oxidation of these inclusions, the waters in the mine become acid. In regions of bituminous coal in the U. S. A. acid drainage waters form in great quantities; they contaminate streams and cause corrosion of mine equipment. In this connection, the oxidation of coal inclusions is of great practical importance, and, in investigating the effect of bacteria on sulfides, special attention has been directed toward it. Since pyrite is the most widespread sulfide in nature, these investigations are undoubtedly of interest as well in understanding the role of bacteria in the oxidation of ore deposits.

Laboratory experiments of Temple and Delchamps (1953), and also of Lethen, Braley, and McIntyre (1953), on the influence of cultures of Thiobacillus ferrooxidans and T. thiooxidans on samples of marcasite and pyrite extracted from coal demonstrate a considerable acceleration of the oxidation of these minerals over the rate in a sterile control. The laboratory experiments have been supported by observations of Temple and Delchamps, who, in studying only exposed coal beds, found the reaction of the water in this zone to the neutral and detected no Thiobacillus ferrooxidans or T. thiooxidans. But after a few days the water acquired high acidity and both species of bacteria were present in large numbers. Ashmeed (1955), studying the formation of acid in the coal mines of Scotland, concluded that four-fifths of all the sulfuric acid that formed was due to the oxidation activity of bacteria. From this it is clear that the role of bacteria in the oxidation of iron sulfide is very great and that it is necessary to consider this factor where the process is effective.

Already, in studying the oxidation of sulfur contained in coal inclusions, it has been discovered that the species Thiobacillus ferrooxidans has the chief role in the oxidation of sulfides. This organism grows exclusively in an acidic environment, at a pH below 4.5. Colmer, Temple, and Hinkle (1949) have shown that this species oxidizes the lower oxides of sulfur as well as of iron. Although the capacity to oxidize sulfur compounds is denied by Leathen and others (Leathen, Kinsel, and Braley, 1956), experiments by a whole series of workers (Bryner, Beck, Davis, and Wilson, 1954; Bryner and Anderson, 1957; Bryner and Jamerson, 1958), conducted on sulfides containing no iron, confirm the view that the organism does possess this capacity. In these experiments, the investigators used cultures isolated from the copper and lead-zinc deposits of Bingham Canyon (Utah, U. S. A.) and Cananea (Sonora, Mexico). The experiments were carried out in percolators, in which ore was placed, mixed with sand, mineral medium, and culture of Thiobacillus ferrooxidans. In control percolators, the oxidation was effected under sterile conditions. In the experiments of Bryner, Beck, and others, in which bacteria acted on chalcopyrite, ten times as much copper went into solution as in the sterile control.

The results of bacterial action on other copper-sulfide minerals are shown in Table 1.

TABLE 1. Biological Oxidation of Sulfide Minerals, in mg of Leached Copper (after Bryner, Beck, Davis, and Wilson, 1954)

Time in days	Covellite		Chalcocite		Bornite		Tetrahedrite	
	expt.	control	expt.	control	expt.	control	expt.	control
7	13.5	4.6	12.0	6.0	15.0	0.8	0.8	0
14	39.0	10.1	23.0	12.8	23.0	2.2	1.7	0
21	79.0	14.1	52.4	19.0	31.0	2.6	2.1	0
28	112.0	17.7	83.4	26.6	38.0	3.9	2.9	0
35	126.0	20.5	91.4	41.4	41.4	4.17	2.9	0
42	134.8	22.4	99.6	39.2	46.1	5.1	2.9	0

One of the most interesting physiological peculiarities of Thiobacillus ferrooxidans, of great value for growth in copper deposits, is its resistance to copper. We isolated this organism from water containing 7.5 g/liter of copper. According to American authors, bacteria adapted to copper tolerate that metal in concentrations of 12 g/liter. It is interesting to note that the capacity to tolerate high concentrations of copper is not merely an adapted feature, but is a property of the whole species. In our experiments, a culture that was obtained from a spring containing no copper was able to grow and oxidize iron, although at a reduced rate, in a medium with a copper concentration of 2 g/liter

Bryner, Anderson, and Jamerson (Bryner and Anderson, 1957; Bryner and Jamerson, 1958) made special studies of the oxidation of molybdenite, which is one of the most stable sulfides. Bryner and Anderson have shown that seven times as much molybdenum goes into solution in the presence of Thiobacillus ferrooxidans as in the sterile control. The addition of pyrite has proved to have a positive effect on the oxidation of molybdenite; when pyrite was added, 30 times the molybdenum went into solution that dissolved in the sterile control.

The mechanism of oxidation is obviously not the same for the various sulfides. The oxidation of copper and molybdenum sulfides, in contrast to pyrite, probably takes place by oxidation of sulfur atoms by bacteria according to the equations:

$$4Cu_2S + 9O_2 = 4CuSO_4 + 2Cu_2O,$$
$$2MoS_2 + 9O_2 + 6H_2O = 2H_2MoO_4 + 4H_2SO_4.$$

Concerning the oxidation of pyrite, Temple and Delchamps (1953) believe this process to occur in nature according to the following scheme

$$1.\ FeS_2 + H_2O + 3^1/_2O_2 \rightarrow FeSO_4 + H_2SO_4.$$

The first stage of oxidation is entirely chemical. Then the ferrous sulfate is converted to ferric sulfate.

$$2.\ 2FeSO_4 + {}^1/_2O_2 + H_2SO_4 \rightarrow Fe_2(SO_4)_3 + H_2O.$$

Since this reaction is almost impossible chemically in an acid environment, the oxidation takes place through the influence of Thiobacillus ferrooxidans. The ferric sulfate reacts with available pyrite, and the sulfur forming from this reaction is oxidized by Thiobacillus ferrooxidans.

$$3.\ FeS_2 + Fe_2(SO_4)_3 \rightarrow 3FeSO_4 + 2S,$$
$$4.\ S + 1\frac{1}{2}O_2 + H_2O \rightarrow 2H + SO_4''.$$

The merit of this scheme is that it includes both the purely chemical oxidation of pyrite and the role of bacteria in the process. It is possible that Thiobacillus ferrooxidans is able to act directly on the pyrite molecule, since the oxidation of the mineral is sharply accelerated in the presence of this organism. In one of our experiments, in which Thiobacillus ferrooxidans acted on samples of coal containing pyrite, 950 mg/liter of ferric oxide formed in three weeks in this environment, whereas only 5 mg/liter formed in the sterile control.

The indirect role of Thiobacillus ferrooxidans, which involves the formation of ferric sulfate (according to reaction 2), is also very large. As we have indicated already, molybdenite oxidized four times as fast in the presence of pyrite as simply in the presence of bacteria; it is clear that this points to an oxidizing role of ferric sulfate, which forms from pyrite under the influence of bacteria.

In our experiments the effect of Thiobacillus ferrooxidans was tested on ore from the Degtyarka deposit (Urals), consisting of pyrite and chalcopyrite. Ground ore was placed in percolators. In the experimental percolator a composite culture was added, isolated from the Krasnogvardeiskii deposit in the Urals; an antiseptic was added to the control percolator.

The results of these experiments are shown in Table 2.

TABLE 2. Oxidation of Pyrite Ore by a Culture of Thiobacillus ferrooxidans

Exptl. period, day	Content of Cu, mg/liter		Content of Fe^{3+}, mg/liter		pH	
	expt.	control	expt.	control	expt.	control
Start of experiment	11	9		3.5	3.9	3.8
7th	30	8.5	9.4	2.7	3.95	4.25
13th	70	13	93	1.5	2.6	3.9
18th	114	14	390	1.3	2.2	3.8
30th	130	16	1150	2.3	2.0	3.8
37th	140	16	1700	8	1.9	3.8

As seen from the data of Table 2, eight times as much copper went into solution in the presence of bacteria as in the sterile environment. The concentration of ferric oxide increased sharply. In order to determine the role of ferric sulfate in the leaching of copper, we set up another experiment with the same ore. A concentration of ferric oxide was prepared in a sterile solution that was equal in concentration to that at the end of the preceding experiment. After five days the concentration of copper in solution reached 75 mg/liter, whereas it had been but 15 mg per liter at the beginning of the experiment; after this the solution of copper ceased. The content of ferric sulfate diminished sharply. From this it may be seen that copper is leached in the presence of ferric sulfate, but that the process quickly stops in a sterile environment, since the iron goes into the ferrous form.

On the basis of the laboratory data we may conclude that the oxidation of sulfide ores is speeded up 5-10 times by the presence of Thiobacillus ferrooxidans. The bacteria act on the sulfides both directly, oxidizing atoms of sulfur, and indirectly, by forming a powerful oxidation agent of ferric sulfate.

The laboratory results cannot be applied unreservedly to the process taking place in nature; the role of bacteria in oxidizing any particular sulfide deposit will depend on a whole series of conditions.

A certain amount of moisture is required for the growth of all bacteria, and for aerobic bacteria, oxygen is also necessary. Consequently, bacterial oxidation may take place to a certain level, where oxygen-bearing ground water penetrates. In this connection, highly permeable rocks are of great importance. Fracturing in the surrounding rocks, especially fracturing in the ore body itself, greatly facilitates the entrance of water and increases the contact surface between water and sulfides; therefore, extensive fracturing creates more favorable conditions for the growth of microorganisms. Climatic conditions, especially temperature and amount of precipitation, should have a considerable influence on the oxidation of ores by bacteria. The activity of any species of bacteria will depend on the factors listed above, and Thiobacillus ferrooxidans, in addition, will be further affected by a number of conditions associated with the physiological peculiarities of this organism. As already noted, Thiobacillus ferrooxidans requires an acid environment for its growth. In this connection, the mineral composition of the ore is of great importance; pyrite is especially important, since when it, a disulfide, is oxidized, large quantities of acid are formed. The nature of the ground water and of the host rocks is also an important factor. If the country rock, such as carbonate, will neutralize the acidity, unfavorable conditions are created for the growth of Thiobacillus ferrooxidans; when pyrite is absent, or for some other reason the water will have a pH that is nearly neutral, the role of this organism will clearly be small. It is possible that, under such circumstances, other species of sulfur-oxidizing bacteria will participate: Thiobacillus thioparus and, if the waters contain nitrates, Thiobacillus denitrificans.

The role of Thiobacillus ferrooxidans in the oxidation of ores was studied by us at several deposits: in the Armenian SSR, in the Urals, and on the Kola Peninsula.

The Akhtala lead-zinc deposit in Armenia may serve as an example of a deposit having favorable conditions for the development of Thiobacillus ferrooxidans. The ore body is near the surface; rain water easily penetrates along cracks and causes considerable wetting of the mass. The water drains off in creeks flowing both from workings and from abandoned tunnels. The pH of the water ranges from 1.4 to 3.9, and, at the same time, has a large iron-oxide content. Since iron sulfate hardly ever changes to the oxide form by chemical means at such low pH, it is clear that the iron has been oxidized by Thiobacillus ferrooxidans, which has been detected in all the water samples. It is likely that the oxidation of the ore is favored by the fact that the deposit has been worked for a long time (since the end of the eighteenth century), and many sections have been exposed for a long time. Thiobacillus ferrooxidans has been found in samples of copper, zinc, and lead ore that have lain on the dumps for nearly a year. Ore that has lain on dumps for several years has become deeply altered; the copper sulfides have been oxidized to sulfosalts, and the rocks have become altered to soft clayey masses. It is probable that Thiobacillus ferrooxidans took part in the oxidation process (Ivanon, Lyalikova, and Kuznetsov, 1958).

We investigated four sulfide deposits in the Middle Urals. Although Thiobacillus ferrooxidans was found in all the deposits, its role in the oxidation process was clearly very different for the various deposits. At the largest of the Middle-Ural sulfide deposits—Degtyarka—Thiobacillus ferrooxidans was detected in great numbers. The number in dumps proved to be 1000-10,000 cells per gram of rock. The organism was found in similar numbers at the 250 level in water from a drill hole passing through the ore body and in water flowing along the drift. It was absent in samples of freshly extracted ore and in water passing through the country rock. We have attempted to compare the data concerning the presence of Thiobacillus ferrooxidans with the ecological conditions. At the Degtyarka deposit these conditions are perfectly favorable for the development of the organism, since there is considerable moisture and the ground water is acid (pH of 1.7-3.7). As indicated above, laboratory experiments have shown that Thiobacillus ferrooxidans can oxidize the Degtyarka ore, using the ore as a source of energy.

Thus, not only is Thiobacillus ferrooxidans present at the Degtyarka deposit, but the ecological conditions and the substratum are favorable for its growth. In considering all these facts, we may conclude that this organism is an important factor in the oxidation of the indicated deposit, and the high acidity and the concentration of copper (exceeding 1 g/liter in the ore water) controls its activity to a considerable extent. An example of the role of Thiobacillus ferrooxidans may be observed at the Krasnogvardeiskii deposit, where the leaching of copper is as great as at the Degtyarka deposit; a special precipitation arrangement is set up to trap the copper that is carried in solution. Spontaneous heating and even spontaneous combustion have been repeatedly observed at the Krasnogvardeiskii deposit. The number of Thiobacillus ferrooxidans in the water passing through zones that are in danger of fire (zones that have been filled with clay pulp in putting out a fire or as a preventive measure) reaches 100,000 cells per milliliter. The possibility should not be excluded that this organism plays a definite role in the initial processes of spontaneous combustion of ore. At the Third International deposit, Thiobacillus ferrooxidans was isolated from samples of fresh, little-oxidized ores. But, although there are individual zones at the deposit where bacterial oxidation occurs, the over-all role of bacteria in the oxidation of this deposit is clearly small. This conclusion is indicated by the small quantity of copper in the mine waters. It is possible that the massive and disseminated ores at the Third International deposit possess great resistance to both chemical and bacterial oxidation.

At the Pyshma deposit the investigated organism is almost completely absent; at least it plays no role in the oxidation of ores at the 55-85 m levels.

The copper-nickel deposits on the Kola Peninsula are situated beyond the Arctic Circle, under completely different climatic conditions from those previously examined. We investigated the Nittis-Kumuzh'e mine and the abandoned Sopcha deposit at Monche-tundra, and also the Kaula and Kammikivi deposits in the Pechenga ore field. In comparison with the Degtyarka and Krasnogvardeiskii deposits, the oxidation of the ores of the copper-nickel deposits on the Kola Peninsula is much weaker. This is indicated by the low content of leached metals in the mine waters. For example, the waters of the Nittis-Kumuzh'e deposit contain 15 mg/liter of nickel and 2 mg/liter of copper. Despite this, Thiobacillus ferrooxidans was found at all deposits except Sopcha. In the waters at Nittis-Kumuzh'e this organism was found in 20 of 23 samples, but in small numbers. In but two samples of acid water was the number as great as 1000 cells per milliliter. A factor unfavorable for the growth of Thiobacillus ferrooxidans is the absence of pyrite (and the consequent lack of proper acidity of the water). The temperature, which is constantly +4° in the mines, is also an unfavorable circumstance for development of bacterial oxidation. But this does not mean that the bacteria are in an inactive state under these conditions. Isolated cultures were able to grow at this temperature, although more slowly than at 20-30°. Besides Thiobacillus ferrooxidans, T. thiooxidans and Gallionella were also found at the deposits on the Kola Peninsula; large numbers of the latter were found growing on glass slides placed in a tunnel. As the example of the investigated deposits shows, Thiobacillus ferrooxidans, and possibly other sulfur-oxidizing bacteria as well, may play an important role in the oxidation of sulfide ores. The role of the bacteria depends on many factors; the study of the bacteria in nature is therefore necessary to connect their function closely with the geology and hydrogeology of the deposit.

The oxidation activity of bacteria is not merely of theoretical interest; it is of great practical significance. We have already mentioned the damage in some parts of the U. S. A. caused by acid waters arising during oxidation of sulfur in coal inclusions (Hodge, 1937). Since the formation of acid waters is not so widespread in the coal mines of this country, there is no necessity of suppressing the oxidation activity of the bacteria. We have turned our attention to another side of the question— to the fact that coal is freed of sulfur by the oxidation of sulfurous inclusions— and we have tried to employ bacteria for enriching the coal. In investigating this problem it has been shown that coal may be freed of 25% of the sulfur in a month by the action of bacteria (Zarubina, Lyalikova, and Shmuk, 1959).

The oxidation of ores by bacteria has long been used for leaching low-grade ore and dumps, although the bacterial action in this process received little attention. Dumps have been leached at the copper deposits of Bingham Canyon in the U. S. A. and at Cananea in Mexico. According to Weed (1956), who described the arrangement for leaching in Mexico, 40,000,000 tons of dump material, with an average copper content of 0.2%, had collected near the mine. The dumps were leached by percolating water through them; the water was then collected in underground reservoirs. The copper was precipitated from the water by an exchange reaction with iron. By the leaching of dumps, a secondary recovery of nearly 650 tons of copper per month was obtained. Bacteria clearly took an active part in this leaching process, since they were isolated by Bryner and Jamerson from this setup; and in laboratory experiments the oxidation of sulfides was greatly accelerated.

Underground artificial leaching is one of the processes for which bacteria may be used. According to Aglitskii and Dyn'kina (1956), from 5 to 20% of the ore remains behind in worked-out deposits; further mining operations are of no value in recovering this, either because the ore mass is too badly shattered or because the content of copper is too low (low-grade ore). To extract copper inaccessible by mining operations, artificial underground leaching is employed, by sending water through specially drilled holes from the workings of the deposit. We employed artificial underground leaching at the Severo-Karpushinskii, Degtyarka, and other deposits. For liquid we used either fresh water or tailings water from the recovery-precipitation arrangement; by far the best results were obtained with the latter. This may be explained by the large quantity of ferric sulfate and sulfuric acid in the tailings water, but there is no doubt that an important role was played by the large number of bacteria. Analyses of the waters from the precipitation arrangement at Krasnogvardeiskii show that Thiobacillus ferrooxidans is present there in numbers of 10,000 cells per milliliter; and in the copper concentrate at the Degtyarka deposit, there are more than 1,000,000 cells per gram. If, during artificial underground leaching, we create the optimum conditions for development of bacteria by using acid water with the addition of nitrogenous salts, the leaching process will probably be greatly accelerated.

Acid solutions of ferric sulfate have long been used for leaching sulfide ores; in this process the metal of the ores goes into solution and the ferric sulfate is reduced. For simple leaching this method has the advantage of being a quick process, but it has a difficulty in the necessity of regenerating the iron that was reduced, and in an acid environment this is a very difficult task by chemical means.

In 1958 the Kennecott Copper Company of the U. S. A. obtained a patent on a method of leaching ore by which the solution of ferrous oxide was regenerated by a culture of the bacteria Thiobacillus ferrooxidans (Zimmerley and others, 1958). Thus, in this method, emphasis is placed chiefly not on direct oxidation of the ore by the bacteria, but on the indirect activity of the bacteria, in the continuous formation of ferric sulfate. This scheme is illustrated diagrammatically in the accompanying figure. The method may be used for leaching copper and zinc ores, and the content of chrome and titanium in low-grade ores of these elements may be increased by leaching of the iron.

There is no doubt that this technique will also find application in our industry, and this activity of bacteria, playing a great role in geological processes, will be turned to the advantage of our national economy.

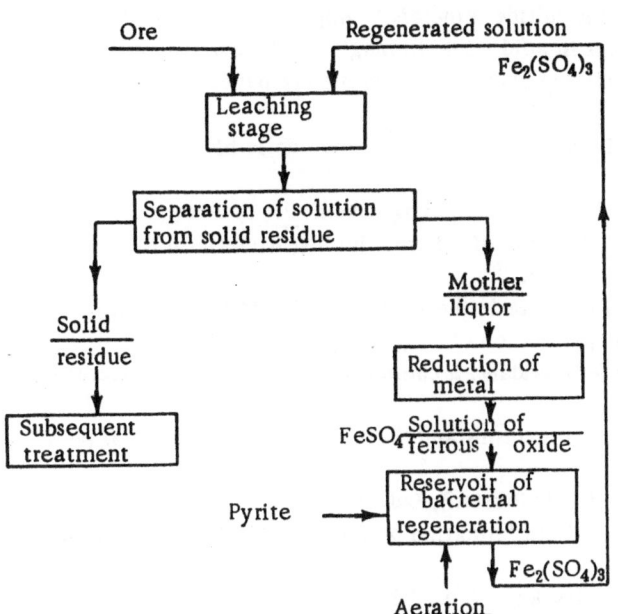

Diagrammatic sketch of the cyclical process of leaching by using bacteria (Zimmerley, 1958).

Conclusions

1. The oxidation of sulfide ores was considered until recently to be purely a chemical process (Smirnov, 1955).

But in recent time there has come to be no doubt that microorganisms participate in this process. The newly discovered organism Thiobacillus ferrooxidans has occupied a prominent place in the study of biological oxidation of sulfides; the organism was first isolated by Colmer and Hinkle from the acid waters draining coal mines.

2. Because acid waters draining coal mines cause considerable damage—producing corrosion—special attention has been given by investigators to the study of oxidation of iron sulfides (pyrite and marcasite) that form inclusions in coal beds. The work of Temple, Delchamps, Leathen, and others has demonstrated the role of Thiobacillus ferrooxidans in the oxidation of these sulfides and in the formation of sulfuric acid, three-quarters of which, according to Ashmeed, is formed by biological means.

3. Laboratory experiments with various sulfide ores (covellite, chalcocite, molybdenite) by Bryner, Beck, Anderson, and others have demonstrated the capacity of Thiobacillus ferrooxidans to oxidize sulfides of other metals as well as of iron. It is clear that biological oxidation is also observed in nature, as indicated by the large number of Thiobacillus ferrooxidans in lead-zinc and pyrite-sulfide deposits.

4. As a result of chemical and biological oxidation, which takes place three to ten times as fast as chemical oxidation alone, the metals that occur in sulfides go into solution. Copper that is dissolved in mine waters is trapped in many mines by precipitation arrangements. In addition, leaching of low-grade ore is done at individual mines, such as in Mexico. The oxidation activity of bacteria is thus put to practical use. By creating better conditions for bacteria to grow, artificial leaching may be improved.

5. Besides the direct effect of Thiobacillus ferrooxidans on the sulfides of various metals, the organism also accelerates oxidation indirectly by converting the ferrous sulfate $FeSO_4$ to the ferric sulfate $Fe_2(SO_4)_3$. This latter compound reacts with sulfides and oxidizes them as it is being reduced. The capacity of Thiobacillus ferrooxidans to oxidize ferrous sulfate is used for leaching ore in the method developed by the Kennecott Copper Co.

6. The intensity of the oxidation activity of bacteria in ore deposits is closely associated with the geology and hydrochemistry of the deposits. There is special significance in the fracturing of an ore body and in the wetting of the deposit.

LITERATURE CITED

Aglitskii, V. A. and Dyn'kina, S. E. 1956. "Underground leaching of copper." Gornyi zhurn., No. 11.
Ashmeed, D. 1955. "The influence of bacteria in the formation of acid mine waters." Colliery Guard., 190, p. 694.

Bryner, L. and Anderson, R. 1957. "Microorganisms in leaching sulfide minerals." Indus. and Engng. Chem. 49, p. 1721.

Bryner, L., Beck, J., Davis, D., and Wilson, D. 1954. "Microorganisms in leaching sulfide minerals." Indus. and Engng. Chem. 46, p. 2587.

Bryner, L., and Jamerson, A. 1958. "Microorganisms in leaching sulfide minerals." Appl. Microbiol., 6, No. 4.

Colmer, A. and Hinkle, M. 1947. "The role of microorganisms in acid mine drainage." Science, 106, p. 252.

Colmer, A., Temple, K., and Hinkle, M. 1949. "An iron-oxidizing bacterium from the drainage of some bituminous coal mines." J. Bact., 59, p. 317.

Helbronner, A. and Rudolfs, W. 1922. "L'attaque des minérals par bactéries. Oxidation de la blende." C. r., 174, p. 1378.

Hodge, W. 1937. "Pollution of streams by coal mine drainage." Indus. and Engng. Chem. 29, p. 1048.

Ivanov, M. V., Lyalikova, N. N., and Kuznetsov, S. I. 1958. "The role of sulfur-oxidizing bacteria in the weathering of rocks and sulfide ores." Izv. AN SSSR, seriya biol., No. 2.

Leathen, W., Braley, S., and McIntyre, L. 1953. "The role of bacteria in the formation of acid from certain sulfuritic constituents associated with bituminous coal." Appl. Microbiol., 1, No. 2.

Leathen, W., Kinsel, N., and Braley, S. 1956. "Ferrobacillus ferrooxidans, a chemosynthetic autotrophic bacterium." J. Bacteriol., 72, No. 5.

Rudolfs, W. 1922. "Oxidation of iron pyrites by sulfur-oxidizing organisms and their use for making mineral phosphates available." Soil. Sci. 14, p. 135.

Smirnov, S. S. 1955. The Zone of Oxidation at Sulfide Deposits [in Russian], Izd. AN SSSR.

Temple, K., and Delchamps, E. 1953. "Autotrophic bacteria and the formation of acid in bituminous coal mines." Appl. Microbiol., 1, p. 255.

Weed, R. 1956. "Cananea's program for leaching in place." Mining Eng., 8, p. 721.

Zarubina, Z. M., Lyalikova, N. N., and Shmuk, E. I. 1959. "An investigation of the microbiological oxidation of pyrity in coal." Izv. AN SSSR OTN. Metallurgiya i toplivo, No. 1.

Zimmerley, S., Wilson, D., and Prater, J. 1958. "Cyclic leaching process employing iron-oxidizing bacteria." US Pat. No. 2829964.

THE APPLICATION OF BACTERIAL METHODS
OF BENEFICIATING ORES OF NONFERROUS METALS

V. I. Ivanov

(Institute of Microbiology, Academy of Sciences, USSR, Moscow)

During seven years of increased production of nonferrous metals, together with intensification of existing methods, a leading role has been assigned to the development and industrial introduction of new, more improved and profitable processes of beneficiating ores and of methods of composite studies of the valuable components from the beneficiation products.

The necessary requirements of nonferrous metals in this country may be met by discovery and utilization of new rich deposits of nonferrous metals (such as the Gai deposit in the Southern Urals), and also by bringing into production known reserves of low-grade ore, dumps of so-called waste rock, tailings of beneficiating plants, and other industrial waste products on the surface of the ground, which, despite the low concentration, contain considerable quantities of metals in critical supply.

Beneficiation experts and metallurgists are faced with the very urgent problem of treating ores that are difficult to beneficiate, such as copper-zinc sulfide ores with cryptocrystalline structure and the carbonate manganese ores of the Urals, the complex composition of which has prevented obtaining high recovery.

It is well known that a solution to the problem of treating low-grade and refractory ores is closely connected with the development of new methods, since existing methods at this stage of technological development are not profitable.

Recently in the U. S. A. new microbiological methods of beneficiation in the hydrometallurgy of copper and zinc have been developed; these methods have grown out of the discovery by microbiologists of a newly identified species of bacteria, Thiobacillus ferrooxidans, in the acid waters of coal mines; this species of bacteria may increase the rate of oxidizing sulfide minerals 10-20 times over ordinary chemical oxidation. Not long ago workers in the U. S. A. also showed that it is basically possible to use bacteriological treatment of low-grade oxidized manganese carbonate ores. Similar studies have also been made in other countries.

According to experiments conducted by American scientists, the principal defect of bacterial leaching is the considerable time involved in the leaching cycle. Even after using air-lift percolators, guaranteeing continuous circulation of the nutrient solutions and insuring weekly replacement of these solutions, the time involved in leaching low-grade copper ores to obtain the equivalent of dump tailings is about 80 days, and to obtain the equivalent of rich copper flotation concentrates, is up to one year.

One might reconcile himself to this long period of leaching if he is considered using spontaneous biological processes in a deposit itself, on dumps of waste rock, or in the tailings of beneficiation plants. But to introduce biological methods into industry as independent methods of treatment, it is necessary to reduce the time of leaching to several days.

The prolonged time of bacterial leaching thus far observed is apparently associated with a retarded stage of solution of the sulfides of heavy nonferrous metals or of manganese carbonate, which are but slightly soluble substances, and is not due to the rate of bacterial oxidation of sulfur (in the sulfide form) or of ferrous oxide.

The increased rate of bacterial solution of copper and molybdenum sulfides in the presence of pyrite may be explained by the increase in solution of the latter because of the aggressive action of ferric sulfate that forms by bacterial oxidation of pyrite.

The Kennecott Copper Company, in taking out a patent on the biological leaching of copper and zinc from low-grade ores, and also on the biological beneficiation of molybdenum, iron-chrome, and iron-titanium concentrates, for intensifying bacterial leaching, has successfully combined the bacterial process with the process, long known by metallurgists, of leaching sulfides by solutions of ferric-oxide salts.

The defect of bacterial leaching, as indicated above, is the long time involved, and the defect of leaching with a solution of ferric-oxide salts is the difficulty of regenerating the solvent: in acid solutions bivalent iron is very difficult to oxidize, even when the material is aerated. The introduction of ferric sulfate into the bacterial medium reduces the time required for leaching, and the introduction of bacterial cultures into a solution of ferrous sulfate increases the rate of oxidation of iron by aeration, and this practically solves the problem of regeneration of the solvent.

This combination of two different processes does not imply that bacterial leaching is replaced by chemical, since bacteria oxidize ferric iron as the solvent is reduced, just as the regeneration of the solvent in the leaching process itself shifts the equilibrium of the reaction in a direction to cause solution of a new batch of sulfides. The leaching of sulfides by ferric sulfate in the presence of bacteria is also accelerated by the oxidation of elemental sulfur, formed during abiogenic chemical leaching, by the bacteria.

There is no doubt that the indicated combination of bacterial and chemical processes brings bacterial methods near the satisfactory point of introduction into industry.

We have dealt with none of the actual data obtained in the investigations at the Kennecott Copper Company or at the Bureau of Miners, which might be necessary for an economic evaluation of bacterial leaching or of the possibility of using the method in treating a particular type of ore. In our country scientific research in this direction has already begun, and it is necessary to expand this work in every way possible.

However, the status we have attained in this field does not signify that a definite industrial and economic evaluation of the biological method of beneficiation should be given after obtaining actual results on a definite series of specific ores. The known data in this field are already sufficient to assure us that the biological method is the most promising technique for treating low-grade and refractory ores of copper-zinc and other deposits, and it deserves the wide attention of microbiologists and engineers.

Careful study of published data leads one to conclude that we stand at the threshold of creative control in microbiological processes in mineral deposits; we are at the point where we may improve the processes and use them as independent methods of treatment at mines, beneficiating plants, and hydrometallurgical factories.

Conclusions

Many nonferrous metals still remain in the tailings of beneficiating plants and in other waste piles of the mining industry.

In the U. S. A. new microbiological methods of hydrometallurgy of copper and zinc have begun to appear.

It is necessary to develop methods of controlling microbiological processes in mineral deposits and to use microbiological methods for treatment at mines and beneficiating plants.

SOME POINTS CONCERNING THE STUDY OF BIOLOGICAL, SULFURIC-ACID WEATHERING

G. I. Karavaiko

("Uralmekhanobr" Scientific-Research and Planning Institute, Sverdlovsk)

We had made no study of the role of sulfur-oxidizing bacteria, particularly Thiobacillus thiooxidans and T. ferrooxidans, in sulfuric-acid weathering in connection with the origin of bauxite deposits, either in swamps or at sulfide deposits, until recently.

Therefore, the present communication is based completely on the literature. Until now studies have been made of the role of the above-indicated bacteria in sulfuric-acid weathering of sulfur deposits and of sulfide ores, and also the role of these organisms in mineral springs, soils, and elsewhere.

Sulfur-oxidizing bacteria play an important part in the sulfur cycle in medicinal muds, soils, mineral springs, and deposits of sulfur and sulfide ores.

According to Ivanov, Lyalikova, and Kuznetsov (1957), Thiobacillus thiooxidans takes an active part in sulfuric-acid weathering of sulfur deposits; and Thiobacillus ferrooxidans, according to the same authors, is instrumental in the oxidation of sulfide ores. This fact is also pointed out in the paper of Temple and Colmer (1954), where it is stated that Thiobacillus ferrooxidans have been found in the acid drainage waters of coal mines.

The data of Kuznetsov (1955) and Emoto (1933) indicate that Thiobacillus thiooxidans and similar species take an active part in the oxidation of molecular sulfur in mineral springs. The wide distribution of Thiobacillus thiooxidans in such springs points to a biogenic origin of the acidity that ensues from oxidation of the sulfur to H_2SO_4. The distribution of sulfuric-acid processes in swamps has been but little studied; however, the high acidity of mangrove swamps in tropical zones and in coastal swamps of temperate climates indicates that the mentioned processes are at work in such places. Furthermore, this process is important because, according to Yakovleva (1959), sulfuric-acid weathering in swamps is a prime factor in the mobilization of alumina. It is therefore necessary to give serious attention to the study of sulfuric-acid weathering in swamps and peat bogs and to the role of sulfur-oxidizing bacteria in this process. It may be stated, judging just from the literature, that here also the biological factor is very important.

The data of Dreves (1928), who succeeded in isolating Thiobacillus thiooxidans from acid swamps soils in Germany, also favor this view.

In investigating sulfuric-acid weathering in swamps in connection with the origin of bauxites, we should call attention to the swamp on the Karelian Isthmus (Yakovleva, 1959), where sulfuric acid is formed by oxidation in a peat bog rich in iron sulfide. The sulfuric acid acts on the alumino-silicate products of weathering that have been brought into the swamp from higher parts of the surrounding slope; it is apparently a factor in the mobilization of free aluminum. The accumulation of the latter takes place in the lower parts of the swamp, where the environment is more alkaline. Here, aluminum accumulates in a peat bog, since peat possess a high capacity of absorption relative to aluminum. Titanium experiences some displacement under these conditions, apparently forming complexes with organic material. Silicon is also removed. Consequently, organic material is of considerable importance in the transfer of aluminum and titanium.

In regard to sulfuric-acid weathering of bed rock in the vicinity of a sulfide deposit, in connection with the formation of bauxites, the matter is clear. Lyalikova has shown that Thiobacillus ferrooxidans plays an important role in the oxidation of sulfide ores (Lyalikova, present volume).[*] Sulfuric acid forms during oxidation of pyrite of hydrothermal origin. In contrast to sulfuric-acid weathering in swamps, here there is extensive removal of aluminum from the weathering zone. The amount of titanium increases relatively because of the removal of other elements, although, in fact, titanium is also removed. When deep-seated waters with a pH of 2.0 reach the surface, aluminum precipitates from sulfuric-acid solutions. It has been detected in the muds of ponds where such waters emerge.

There is considerable interest in the efflorescences of salt. These form by evaporation of acid ground water from the zone of sulfuric-acid weathering. Efflorescences of salt attest, as Yakovleva's (1959) analyses have shown, not only to the mobility of aluminum and silicon under these conditions, but also to the mobility of titanium. These investigations were carried on at deposits in the Southern Urals.

From the previous discussion it follows that sulfuric-acid processes play an important role in the geochemistry of aluminum, titanium, and other elements. According to Vlodavets (1926), considerable quantities of titanium have been detected in ground water of sulfur domes, the water of which contain 5% sulfuric acid. Consequently, titanium is in solution in a highly acid environment. This is also borne out by the data of Yakovleva, who has shown that at a pH below 1.5 all elements, including even the most inert titanum, are found in solution. At higher values of pH, up to 4.5, Al and Ti precipitate in the sediment.

Conclusions

Sulfuric-acid weathering plays an important role in bauxite deposits, and, apparently, the role of sulfur-oxidizing bacteria in this process is rather large.

It is necessary to study the role of bacteria that oxidize sulfur and sulfides to sulfuric acid, in sulfuric-acid weathering of swamps, in connection with the origin of bauxite deposits.

LITERATURE CITED

Dreves, D. K. 1928. "Mikrobiologische Untersuchung eines stark sauren Moorbodens." Zentr. f. Bact., Abt. II, B. II No. 76.

Emoto, Y. 1933. "Studien über die Physiologie der Schwefel oxidierenden Bacterien." Bot. Magaz. Tokyo, 47, p. 405, 567.

Ivanov, M. V., Lyalikova, N. N., and Kuznetsov, S. I. 1957. "The role of sulfur-oxidizing bacteria in the weathering of rocks and sulfide ores." Izv. AN SSSR, seriya biol., March-April, No. 2.

Kuznetsov, S. I. 1955. "Microorganisms in the hot springs of Kamchatka." Trudy INMI, 4.

Temple, K. L. and Colmer, A. R. 1954. "Drainage from bituminous coal mines." Chem. Abstr., 48, No. 11.

Vlodavets, V. I. 1926. "Data on the chemical study of the formation of sulfur domes in the Kara-Kum Desert." from the Collection: Sulfur [in Russian], KEPS, Leningrad.

Yakovleva, M. N. 1959. "The geochemistry of Al, Ti, Fe, and Si in sulfuric-acid weathering (in connection with the origin of bauxites)." from the Collection: Bauxites, Their Mineralogy and Origin [in Russian], Otd. geologeogr. nauk AN SSSR, Moscow.

---

[*] See p. 102.